Sporting Times

DOI: 10.1057/9781137275363

Other Palgrave Pivot titles

G. Douglas Atkins: **T.S. Eliot Materialized: Literal Meaning and Embodied Truth**

Martin Barker: **Live To Your Local Cinema: The Remarkable Rise of Livecasting**

Michael Bennett: **Narrating the Past through Theatre: Four Crucial Texts**

Arthur Asa Berger: **Media, Myth, and Society**

Hamid Dabashi: **Being a Muslim in the World**

David Elliott: **Fukushima: Impacts and Implications**

Milton J. Esman: **The Emerging American Garrison State**

Kelly Forrest: **Moments, Attachment and Formations of Selfhood: Dancing with Now**

Steve Fuller: **Preparing for Life in Humanity 2.0**

Ioannis N. Grigoriadis: **Instilling Religion in Greek and Turkish Nationalism: A "Sacred Synthesis"**

Jonathan Hart: **Textual Imitation: Making and Seeing in Literature**

Akira Iriye: **Global and Transnational History: The Past, Present, and Future**

Mikael Klintman: **Citizen-Consumers and Evolutionary Theory: Reducing Environmental Harm through Our Social Motivation**

Helen Jefferson Lenskyj: **Gender Politics and the Olympic Industry**

Christos Lynteris: **The Spirit of Selflessness in Maoist China: Socialist Medicine and the New Man**

Ekpen James Omonbude: **Cross-border Oil and Gas Pipelines and the Role of the Transit Country: Economics, Challenges, and Solutions**

William F. Pinar: **Curriculum Studies in the United States: Present Circumstances, Intellectual Histories**

Henry Rosemont, Jr.: **A Reader's Companion to the Confucian *Analects***

Kazuhiko Togo (*editor*): **Japan and Reconciliation in Post-war Asia: The Murayama Statement and Its Implications**

Joel Wainwright: **Geopiracy: Oaxaca, Militant Empiricism, and Geographical Thought**

DOI: 10.1057/9781137275363

palgrave▸pivot

Sporting Times

Kath Woodward
Open University, UK

palgrave
macmillan

DOI: 10.1057/9781137275363

First published 2013 by
PALGRAVE MACMILLAN

Palgrave Macmillan in the UK is an imprint of Macmillan Publishers Limited, registered in England, company number 785998, of Houndmills, Basingstoke, Hampshire RG21 6XS.

Palgrave Macmillan in the US is a division of St Martin's Press LLC, 175 Fifth Avenue, New York, NY 10010.

Palgrave Macmillan is the global academic imprint of the above companies and has companies and representatives throughout the world.

Palgrave® and Macmillan® are registered trademarks in the United States, the United Kingdom, Europe and other countries.

ISBN: 978-1-137-27537-0 EPUB
ISBN: 978-1-137-27536-3 PDF
ISBN: 978-1-137-27535-6 Hardback

A catalogue record for this book is available from the British Library.

A catalog record for this book is available from the Library of Congress.

www.palgrave.com/pivot

DOI: 10.1057/9781137275363

To Steve

DOI: 10.1057/9781137275363

Contents

DOI: 10.1057/9781137275363

Acknowledgements

Thanks to Philippa Grand at Palgrave for all her support and encouragement in this race against time and thanks to Kerry Lawrence at the Open University for all her technical support and patience. I am so grateful to Steve for enjoying the games with me and for helping out in so many ways, including the Index. Also, I would like to thank Vidhya and her team at Newgen for all their help during the production process.

palgrave▸pivot

www.palgrave.com/pivot

DOI: 10.1057/9781137275363

1
Out of the Blocks: Starting Times

Abstract: *Chapter 1 sets the agenda for the discussion in the book by raising questions about the primacy of time and temporality in sport, about which times matter and about how time matters. This chapter poses some of the questions which are explored in the book by suggesting that sport is in some ways a distinctive field of inquiry but one in which a discussion of time and the conceptualisation of temporality have theoretical importance and resonance across a wide range of social and cultural areas. The immediacy of sport and its immersion in the 'now' generates ways of thinking about the relationships between past, present and future.*

Woodward, Kath. *Sporting Times*. Basingstoke: Palgrave Macmillan, 2013. DOI: 10.1057/9781137275363.

We live in sporting times, not least at the time of writing, in the year of the 2012 Olympic Games and a whole range of other mega sporting spectacles including the men's football European championships, cricket test series as well as all the other regular international competitions, such as Wimbledon, the US Open, the Ryder Cup and golf majors, the NFL Superbowl, basketball and baseball competitions to mention but a few. However, the Olympics, being the biggest and the 'greatest show on earth', is the trigger for this book which explores time in sport and sporting times in order to see what sport can offer to an understanding of time and temporality.

In the UK, especially, 2012 is the year of sporting times not only because this is the year that the games came to London but also because the Olympics and the Paralympics have generated excitement and massive interest, and the athletes who have taken part have become heroic figures in the public arena. It has all worked way beyond the dreams of the organisers and the scepticism of many critics and political activists. The legacy of the games remains uncertain but the duration of the events, that is the real time when it has all been happening, has been an undisputed success in delivering wide engagement and spectacular support, beyond the sale of tickets and interest of sports fans. It is difficult to describe the intensities of the moment without resorting to the familiar hyperbole of sports' commentary with its rhetoric of amazement and disbelief. Each athlete in the post-event interview declared that the experience had been incredible, amazing or unbelievable, or even all three. Superlatives are limited – why not? What else could anyone say? The experience, representation and expression of London 2012 have been remarkable in its democratic reach and the engagement even of sceptics as well as the expected spectacles of the display. The Paralympics and its athletes have generated as much interest and support as has the Olympics which preceded them; the so-often socially inflected embodied differences have become ordinary and pass unremarked. What matters is playing the game. The measure of sporting achievement has been through embodied activities in the field, in the ring, on the track and in the pool in a convivial mix of people who have in common athletic capabilities. The year 2012 has demonstrated well Paul Gilroy's concept of conviviality (Gilroy, 2004) not only in its mixture of democratic participation with the endurance of inequalities, not least the economic differences between the countries participating in the games, but also in the balance between the negative and the positive dimensions of conviviality.

DOI: 10.1057/9781137275363

London 2012 has offered a celebration of the ordinariness of many of the differences between and among people and downplayed the markers of inequality (Stewart, 2007).This is not to say that the games themselves were not marked by inequalities, which also operated routinely (Unequal Time, Cleaners, 2012).

One of the most dramatic demonstrations of the ordinariness of differences, which makes for social inclusion for me in 2012, is the women's boxing with Nicola Adams's wonderful achievement, not 'just' in winning gold but for making women's boxing ordinary as well as spectacular; her great performance in the ring was assessed by the same criteria as a professional men's heavyweight fight in Madison Square Garden. Her speed and agility is resonant of Muhammad Ali's in his prime, as was recognised in the reporting and commentary. What matters is that women's boxing is taken seriously; Adams' skill was acknowledged and not subjected to the trivialisation or sexualisation which so often marks any discussion of women's boxing. This was categorised as indicative of 2012 being 'women's time' or more usually 'girls' time'. The discourse of time includes notions of 'the right time', his or her time or even our time when referring to team GB in the partisan world of sports reporting and in the UK press and BBC commentary. Women's participation and achievement is ordinary in this respect even though the performance in this case, as in so many other instances in 2012, was spectacular and executed with stunning levels of skill and competence, speed and agility. The sport made it spectacular, not the enactment by someone from a marginalised or socially excluded group.

This book, *Sporting Times*, uses the games and the experiences of 'being in time' in these days of sporting wonder to develop some understanding of what is distinctive about how time is lived and played out in sport and how temporality in this context contributes to a wider understanding. Chapter 1 sets the agenda for the discussion in the book by raising questions about the primacy of time and temporality in sport, about which times matter and about how time matters. Sport is often classified as a leisure time activity in its routine enactments and spectatorship. It is time marked-off from employment and paid work, although for many people sport is paid work, not only for professional athletes but also for the huge number of people involved in sports industries, such as support systems and the media: for them sport is working time. However, times and the allocation and availability of time are unequally distributed. Some sports take a long time; some people do not have the time. Sport is all about

DOI: 10.1057/9781137275363

measurement of time, whether in the setting of records, measurements of embodied practices or in the memories of past achievements and failures through which sporting identifications are made.

Time also means immediacy in sport (Rowe, 2008) as records of 'best time', 'personal best' and the 'right time' make for success and reward, whether gained by financial rewards or through personal satisfaction and achievement. Success is condensed into a moment such as at the point of crossing the winning line first or beating a record. Success is measured by achieving a place in history. When British athlete Mo Farah added the gold of the 5,000m to his 10,000m gold medals he was called 'The History Man' as he is only the seventh man in history to have won gold in both competitions (Greenslade, 2012). Time, especially the rapid completion of a task, makes history, that is, the inclusion of timed records into recorded historical time. This is why through a dialogue with the past, which informs our understandings of the present, both past and present are projected into the future. As Eviatar Zerubavel argues, to make sense of now we need to understand how we envision the past (Zerubavel, 2004) because the stories we tell ourselves to make sense of where we are now are constructed in the present.

In sport, history takes shape through memory in specific ways, but is nonetheless as political and imbued with inequalities as well as opportunities as the more explicitly controversial elements of Zerubavel's time maps, which demonstrate the particularities of memory in different spatial contexts. The Olympics bring together over 204 nations across the world in a global mix which maps out new territories and also carries the distinguishing characteristics of the nation states and their histories.

Sport is not only about play; nor are sporting times only records of embodied performances on the track, in the field, in the ring or pool. How sporting histories are organised into coherent stories is both informative and political. Time metaphors as well as the measurement of time play key roles in the evaluation of sport and commentary.

The experience of time is characterised by inevitability and the forces of finitude which characterise organic life and have so concerned humanity as integral to the human condition. The time stories and myths which Barbara Adam recounts are often constructed around human attempts to counter the ravages of time and of death as the inevitable end point of life (Adam, 1990, 1995, 2004, Dreyfus, 1875). Newtonian physics suggests that time flows uniformly without reference to anything else (Newton, 1995 [1686–7]), but perception of time and how time is experienced are clearly social too. Sporting practices are also directed at overcoming some of the limitations of the flesh and the inevitable process of aging. Sport's histories, myths and legends express the endeavours and successes, on occasion, of overcoming some of the constraints of time and flesh and the inextricable interconnections of bodies in time with time.

The concept of time is central to sport, its cultures and practices. *Sporting Times* was the title of a newspaper published from 1865 to 1932, devoted to sports, mainly horse racing – a particularly important sport at the time – which reflected both the relevance and the pleasures of sporting times and the time spent in sport. Sporting times are the times we live in and the media play a key role in this, but there is much more to sporting time and times. Sport is an empirical field which offers particular insights into understanding temporality because of sports' mix of infrastructures of power with strong personal and collective feelings and attachments, all of which are entangled in the immediacy of now. This book uses sport as a focus for the exploration of the conceptualisation of temporality not only by considering the range of ways in which time is important in sport but also by looking at connections and disruptions between these different times. Thus the book aims to:

- unpack time and times as taken-for-granted but under-researched features of sport;
- offer some evaluation and exploration of theoretical approaches to temporality, some of which have been developed in feminist work on temporality and community and the promise of changing times, in relation to temporalities in sport;
- bring together different dimensions of sport which are under-represented in the literature of sport in order to demonstrate how inequalities and opportunities have temporal as well as material spatial dimensions;
- provide an engaging and accessible approach which uses what is particular to sport as well as how sport is part of the wider social-cultural field to draw attention to the politics and culture of time in sport and sporting times;
- provide new ways of thinking about the social and political importance of time through an innovative methodology which relates present, past and future through the process of writing in 'real time', during the Olympics 2012.

DOI: 10.1057/9781137275363

'Real time' – now

I am writing this book in August and September 2012 in 'real time'; a statement which has attracted cries of derision from some scholarly colleagues. What is 'real time'? The very label suggests an authenticity that any serious studies of temporalities would challenge. Writing in 'real time' means writing while watching and is feasible only when watching at a distance and when it necessarily involves some reflection. What is written during the event is like memory re-worked in the writing up through processes of review and editing. It would be difficult to write anything at all during the 100 metres, usually in barely 10 seconds, and it was. I did however try. Time is experience and time is experienced (Adam, 1994) and in this instance my application of 'real time' is an endeavour to capture the intensities of sporting time and times by watching, writing and reflecting during the time span of the Olympics and Paralympics 2012. The suggestion that temporal being (Adam, 1990, 1995, 2004) in this context could be 'real time' invokes some of the possibilities of the iterative and contemporaneous capacities of time. These processes are highlighted by the contradictory nature of the flow of time and the connections and disjunctions between perception and the material objects of perception (Adam, 2004, Hegel (1970 [1840]), Merleau-Ponty, 1968, Zerubavel, 1981).

Writing in 'real time' is a dialogic process in which the experience of time and its affects and sensations are paramount, none more so than in sport. 'Real time' means I too, just like the athletes competing in the games, am writing to the clock: 'real time' has deadlines for authors. 'Real time' includes clock time with intricate and ever more precise mechanisms and technologies of measurement (Frank, 2011). Measured clock time in sport is explicit and visible as an outcome rather than concealed or assumed in the rhythms of temporality (Zerubavel, 1981, 1985). The precise measurement of time is, however, only one strand in 'real time' and in sporting times (Adam, 2004).

'Real time' is now but now necessarily draws upon what is past and is projected into the future. What I write now is meaningful only in relation to the past. In sporting terms, for example, records are broken only because they exceed earlier recorded achievements. Now becomes then and what is written now is communicated in the future. By the time this book is published, some records set at the London games may have been broken and set again. It will, of course, be some time before the much vaunted legacy of the Olympic Games could be realised. In this

DOI: 10.1057/9781137275363

instance, time remains vague and it is not clear at what point any assessment could be made. An evaluation of legacy requires time for reflection and, of course, what is now becomes what has past and there has to be an assessment of the achievements (or failure to achieve) during the intervening years and of what endures over time, however inexact the period of time may actually be. In the contemporary world, time demands attention which means measurement, or at least we think it does. As Barbara Adam argues, modern times are not lived in accordance with the seasons (Adam, 2004) but as quantifiable units. If time is limited, which is a mantra of contemporary life where being too busy to stop and think is a *sine qua non* of being wanted and being successful and of being in time and tune with the times, we have to manage time; time management is a recognised skill and one which can be taught and learned. Sport as play, despite being all about speed and measuring time and creating new records which are even faster than the last, does offer some time out. Another aspect of temporality in contemporary life is its associations not only with pressure but also with risk and insecurity (Giddens, 1989, 1990).

Writing this book during the 2012 games means that I am immersed in the ebb and flow of the intensities and experiences of time and its routine and irregular rhythms. The intensity of the moment and the spontaneity of feeling are characteristics of sport, sport attachments and spectatorship. The experience highlights the association between time and the experiential which characterise phenomenological accounts (Merleau-Ponty, 1962, 1963, 1964, 1968, 1970). These are experiences which raise questions about how time and experience are enmeshed and about the centrality of time in configuring experience. Temporalities also call into question some of the claims of phenomenology and, methodologically, especially the prioritising of the experiential as a means of explaining the importance of time in sport which Merleau-Ponty acknowledges in his later work (Merleau-Ponty, 1968).

Just as athletes have to pace themselves, so do spectators, followers, commentators and analysts. Time, affect, emotion and energy are entangled in the event and the experience; the movement of time also contributes to the generation of excitement. It is not only the breathlessness of physical exertion which evidences the effort and exhilaration of sporting activity. Time and its demands and pressures also create the anticipation and aspiration before the activity and in its duration and in many cases the whole embodied enterprise is directed at meeting the challenge of

DOI: 10.1057/9781137275363

time. Sporting activities are tiring for all who are involved. Just as the energy of the athlete is finite, so too might be the enthusiasm of the spectators and followers. All are caught up in the temporal processes of the event, albeit in different ways, with differentially weighted corporeal involvement and competences. What is common to all is the operation of time. Time is what makes excitement possible, but heightened engagement can be sustained only for a limited period of time. In 2012, however, there seems to have been no reduction of commitment and enthusiasm for the games as the Paralympics followed the Olympics.

Would we be able to sustain this level of enthusiasm for much longer? The media apparatuses which communicate and transmit the embodied practices, routines, spectacles and rituals of sport cannot be financed for limitless periods of time. Is it the passage of time which creates exhaustion and enthusiasm fatigue among spectators or is it the excess of hyperbole and the volubility of sports commentary and media coverage? In the case of the BBC coverage of the 2012 summer games, the explosion of actions, activities and events described as 'amazing', 'incredible' or 'unbelievable', in a manner that is not atypical of contemporary sports journalism itself, calls into question how 'unbelievable' such events and achievements actually might be if the excess had continued much longer.

Time does not, however, simply march on unaffected by social and cultural forces (Adam, 2004, Gale, 1978). Time has capacities of materiality and malleability even if it also carries the inevitability of passing. The unstoppable force of time and the pressure to 'beat the clock' or somehow attain immortality, or at least halt the decline and disintegration of the flesh with the passage of time, with some certainty means mortality for the human body and partly accounts for the range and extent of the 'knowledge practices shared by so many different cultures: to prehistoric myths and rituals, ancient stories and theories concerned with human origin and destiny' (Adam, 2004: 4). Legends become past time, but they are used to reinforce the moment of now and 'real time' in order to attribute substance to that which is so difficult to capture.

The obsession of contemporary sport to achieve records is not so very different from much earlier attempts to construct heroic narratives which defy the constraints of time. Humans are largely left with legends, whereas the non-human participants in sport, especially horses, can hope to achieve immortality through the transmission of their genetic capacities. A three-year-old bay colt named Frankel, at the time of writing, is classified as the best thoroughbred race horse in the world

DOI: 10.1057/9781137275363

(Frankel, 2012) and can expect to earn massive stud fees of £100 million (Forbes, 2012). Horses and dogs can be made sporting heroes through heroic legend and myth, but their genetic makeup is also recognised and can not only be more lucrative (albeit to their human owner) in the short term but also create legends that have embodied actuality in the future.

The experience of 'real time' is not, as the poet Wordsworth described poetry in his Preface to the *Lyrical Ballads*, 'the spontaneous overflow of powerful feelings; it takes its origin from emotion recollected in tranquillity' ([1800] 2006: 151) because the expression of those powerful feelings in the moment of 'real time' refuses reflection and demands an immediacy, not to mention a richness of language which most athletes and commentators lack. The expression of feeling for athletes is largely corporeal and affect is enfleshed (Woodward, 2012). In sport the viewer, listener and spectator is battered by the repetition of hyperbole, which presents yet another dimension of time and a characteristic of temporality, which requires acknowledgement and understanding of the processes of sensation and the to and fro of affects and emotion in what is often an intense sensory experience which constitutes 'real time' and live sport.

'Real time' is not all intensity of course, but it is immediate and sport at the elite level of the games generates excitement that may be lacking in other less-competent sporting activities or may not be a feature of that particular sport. Test cricket has long periods of seeming inactivity. The passage of time has peaks and troughs (Zerubavel, 1981), especially in sport, which is characterised by highs and lows as well as constant, repetitive reiteration. Sport can be routine and even boring at times, but it also manifests the highs of amazing achievements, which may be constitutive of exhaustion in those participating who are physically exhausted and even of those watching. You can have too much excitement and too much hyperbole. This usefully demonstrates another particular feature of temporality which is distinctive from the ebb and flow and the rhythms of temporality which are resonant of music time with its rhythm, tempo and cadences and the cultural practices and choreography of music with its history of synchronisation and commemoration of mythical moments (Eliade, 1989). Sport has rhythms like music. These rhythms, however, are emotive and extreme but also exhausting. Music can be used in sport to elicit emotive affects. The emotive strains of Vangelis's Chariots of Fire signature tune were well chosen in 2012, but then repetition, especially following Rowan Atkinson's reprise of Mr Bean's subversion of its heroic connotations in the Opening Ceremony, did begin to detract from its power and affect.

DOI: 10.1057/9781137275363

These rhythms demonstrate the relationalities of time, none more so than in sport where there is always the tension between winning and losing but also the emotional affective dimensions of watching and of being in time with the experiences of elation and disappointment. Sporting times are all about relationships, but they also pose puzzles about how these relationships work; whatever the apparent certainties of the measurement of time in sport, being in time is also puzzling. Time is condensed and time is transcendent; it is linear and can seem timeless, even during the 100m when medal-winning athletes report sensations of timelessness even within a period of time that is barely 10 seconds (Olympic Stories, 100m, 2012). (Usain Bolt's 2012 Olympic 100m record was 9.63 seconds.) This is because time is closely imbricated with experience and sport offers very particular versions of being in time. Athletes in the 100m even report of being 'lost in time' (BBC Faster Higher Stronger, 2012). It is hard to imagine athletes' perception of the process as timeless but this presents an interesting take on the experience of time where the measurement of time technologies and the human perception of the process might be at variance. These perceptions also give credence to an understanding of specific ways of being in time which could be understood as 'being in the zone' (Csikszentmihalyi, 1975). Being in the zone is a concept which is often invoked in sport and offers one means of engaging with both the timelessness and the transcendence of the experience. The concept of being in the zone raises questions about time being timeless in the perceptions of those who are in time. Perception and experience are central to an understanding of time and temporality which have particular manifestations in sport.

Another element of sporting time which is distinctive is the relationship between time, being in time and flesh; motion is embodied and it is through corporeal effort that athletes both move themselves and propel the objects that are deployed in sport such as javelins, bats and balls, as well as relating to structures and, in some sports, to animals, notably in the games to horses. Boundaries are blurred and elite athletes and non-human components of the sport can become indistinguishable in time. Bodies are however central to the processes of movement in time. Bodies move in space and through time but are also part of events in which objects, sensations and affects are all caught up in time which may or may not constitute a unity of time. Bodies also provide a focus for exploring the inevitability of the passing of time. The degeneration and decline of the flesh is closely imbricated in the notion that time imposes

DOI: 10.1057/9781137275363

restrictions and, of course, finality on flesh. Sport offers some resistance. Nonetheless sporting times also reflect and constitute power relations and differentials themselves. It is not an equal playing field.

Some of these time puzzles are the concern of this book, especially regarding the relationship between past, present and future and how these are conceptualised and the nature of the relationship between inner worlds and the objects of perception; between seer and seen (Mead, 1932). These issues raise questions about how memory, which plays so large a part in the culture of sport, works and whether there are distinctive processes in the constitution of memory in sport which may be more fluid and troubling than the record books suggest. Memory is also constitutive of history and poses questions about whose history matters and what is or is not included in the historical accounts; what is hidden from history as well as which histories matter and how are they made.

The next chapter, Chapter 2, *Time and Motion: Methodologies and Methods*, focuses upon time and times by looking at movement; the movements of bodies which are, of course, central to sport; and chronology, the concept of linear time and the relationship between present, past and future. Sport is concerned with all sorts of movement which are linked to change and which presents problems and challenges for the researcher, especially in terms of approaches to investigating what is moving and changing. Speed matters in sport, not least in the 100m sprint which makes particular demands on the researcher who is in 'real time'. Sport is also news as exemplified through the media coverage of sport which engages with the now as well as discussing the now in relation to past performance and predicting future form in movement across time. This chapter draws upon the experience of writing in 'real time' and starts with sport, in particular the 2012 Olympics and Paralympics, to see what such a study can generate in order to understand some of the key dimensions and processes of being in time and of temporality, that is the conceptualisation of time. Rather than starting with theories of temporality, this book starts with an immersion in the field to see what elements of temporality emerge. The approach takes sport as the starting point, rather than starting with theories of temporality which are then applied to sport. Thus this book seeks to pick out where and how time matters in sport to see where it takes us in understanding temporality.

Chapter 3, *Memories*, explores the to and fro of remembering and re-remembering which are so frequently invoked in sport, whether in

DOI: 10.1057/9781137275363

commentary, following and fandom or in the enfleshed experiences of training which is so often informed by the observation and rehearsal of previous performances. In the games, past records are re-visited and re-invoked in the run up and in the coverage of the events. What has happened in the past is part of what is happening now. We need the past in order to understand the present (Lenskyj and Wagg 2012, Goldblatt and Acton, 2012). This chapter looks at what is and what is not remembered; how these memories are made and re-made, perceived and expressed; and what is involved in the processes of making memories. Memories are reproduced through the narratives which pervade sport in all media; however this chapter draws upon Bergsonian understandings of the movement of memory to challenge the linear chronology of sports story-telling and to look at who is silent in the stories and who speaks and is spoken about. Sporting time is social (Adam, 2004, Gosden, 1994, Gale, 1978), political and historical too (Zerubavel, 2004). Sporting memories are profoundly connected to social relations including social exclusion, notably in terms of those of sex gender (Woodward, 2012a), which has presented an area often marked by silences, invisibility and absences of women in sport, but which is also part of changing times; the excitement and exhilaration of women's events from rowing to boxing in 2012 show the promise of changing times so well. I use the concept of sex gender to embrace the materialities of sex and the capacities of flesh along with the social and cultural forces more familiarly associated with gender in order to sustain a critical analysis which is attentive to the persistence of inequalities and which is informed by a politics of difference (Woodward and Woodward, 2009). Change is built into the processes of making memories, which bring together past, present and future times.

Chapter 4, *Measuring Time; Recording Time*, explores the impact of technologies of measurement and the enfleshed experience of time and of losing time; embodied time and recorded time. It explores questions about how the experience of being in time can be recorded or accessed and how different systems apply to particular sports. The chapter examines some of the ways in which the movement of time is recorded in media coverage – press and news reports – as well as in the records of time measuring devices. Technological advances have led to an even more precise and sophisticated measurement of time which has influenced and shaped the techniques of athletes and their training regimes, strategies and body practices in the Olympics and Paralympics; body practices and recording mechanisms interrelate (Elias, 1992).

DOI: 10.1057/9781137275363

Women's boxing in the Olympics in 2012 offers a most dynamic set of processes in which temporal forces are implicated in diverse ways; being in time and making history by changing times at the right time and where time is both a constraint and a liberating force. Boxing also presents an interesting example of the entangled forces of flesh and time and the measurement of time by the clock and through, for example, comparisons of the women's and the men's events in terms of times and timing. This exploration of boxing leads into the next chapter which engages with future time.

Chapter 5, *Future Time*, looks at how the future and its promises are lived in the present in sport and at how the future relates to the present and the past in relation to sporting performances. Futures are explored in relation to predictions and possibilities and specifically as they involve interventions and experiences, which include the affects on performance and participation of generation and aging, and of impairment as well as social and cultural forces through which inequalities are forged. There are enfleshed restrictions of performance that constrain the promise of futures as well as technologies which create new opportunities, as in the Paralympics. Power and agency underpin these debates. The 2012 Olympics were at various points described as 'women's time' and, in the context of the Spice girls' reprise of their first hit Wannabe which purportedly introduced the notion of girl power, at the Closing Ceremony, the 'girls' games'(sic, more girls than women). The enormous success of women's sporting activities at the games and the operation of different power mechanisms based on divisions of sex gender in sport make the concept of 'women's time' an interesting political device for exploring temporality, especially in the field of sport, but with much wider application especially in relation to the concept of becoming (Deleuze, 1989, Deleuze and Guattari, 1987,Grosz, 2005) which engages with futurity: future time can be conceived outside and beyond the limits of prediction and utopian futures.

The final chapter, Chapter 6, *Changing Times and Changing Time*, uses the example of the games to evaluate the extent to which the regulatory frameworks of sport can create possibilities and the promise of change as well as reinstating constraints and fixities and broadens the scope of the analysis by arguing that sport not only reflects understanding of temporality but also generates ways of being in time and of conceptualising the processes. In spite of new achievements and records which make sport a constantly changing field there are endurances and entrenched

DOI: 10.1057/9781137275363

stabilities, but the book has suggested new ways of looking at time and the possibilities of being in time as well as competing against time. Sport is also part of historical time and sport makes what comes to be history, and making history is another distinctive feature of the ways in which sport is being in time. Chapter 6 ends with an example from the games to show that beginnings and endings are in conversation in making sporting times which might go beyond the discursive limits of time.

The next chapter starts with some of the puzzles identified in this chapter in relation to time and motion and to some of the methodological questions about being in time which this project of writing in 'real time' poses.

DOI: 10.1057/9781137275363

2
Time and Motion: Methodologies and Methods

Abstract: *Chapter 2 focuses upon time and times by looking at movement – the movements of bodies, which are, of course, central to sport – and at chronology and the relationship between present, past and future. Sport is concerned with all sorts of movement which are linked to change, which presents problems and challenges for the researcher, especially in terms of approaches to investigating what is moving and changing. Sport is also news as exemplified through sporting times through which the media coverage of sport engages with the now, as well as discussing the now in relation to past performance and predicting future form in movement across time. This chapter draws upon the experience of writing in 'real time' and starts with sport, in particular the 2012 Olympics, to see what such a study can generate in order to understand some of the key dimensions and processes of being in time and of temporality, that is the conceptualisation of time. Rather than starting with theories of temporality, this book starts with immersion in the field to see what elements of temporality emerge. The approach takes sport as the starting point, rather than starting with theories of temporality which are then applied to sport. Thus* Sporting Times *seeks to pick out where and how time matters in sport and to see where it takes us in understanding temporality.*

Woodward, Kath. *Sporting Times*. Basingstoke: Palgrave Macmillan, 2013. DOI: 10.1057/9781137275363.

The experience and practice of writing this book is itself a part of the methodological approach to temporalities which I have adopted. As suggested in Chapter 1, the empirical trigger to this book is the 2012 Olympics, which are used as the starting point for this exploration of time and times in sport. Writing in what I have called 'real time', that is writing the examples as I watched in most cases, is not used to argue for any greater authenticity of the present but rather to take sporting practices themselves as my starting point and to endeavour to capture something of the element of time which is 'now'. Now is not only the starting point, now *is* the point. The focus on 'real time' is itself something that emerged from the experience of writing the book. From being initially a device for starting the project, the idea of 'real time' became a means of grappling with some of the puzzles which the experiences of being in time and the concept of temporality generate.

Now at speed; the 100 metres

'Now' on 4 August 2012 is the quarter finals of the 100m at 12.20pm BST. I want Usain Bolt, the Jamaican world champion, to win, as do many of those watching, and I also wonder whether he will break a record – his record. I want him to qualify. I know he will, but it is still a tense moment. I hold my breath at the start. Bolt is not known for being quick out of the blocks. Anxiety is increased because the race, spectators and fans are haunted by the memory of Bolt's disqualification from the 100m after a false start in the final of the men's 100m at the World Athletics Championships in Daegu in 2011, following the International Association of Athletics Federations' (IAAF) decision in 2010 to disqualify athletes after a single false start. Could Bolt have another false start? No! He's out of the blocks, not fast; it's relatively slow, for even Bolt, but I know he can win because he decelerates more slowly than everyone else. This is counter-intuitive as I am convinced he is speeding up at the end of the race. I have the knowledge about deceleration rates but I cannot believe it as I watch. What we want to think and what we see is Bolt going faster than everyone else in the last segment of the race. Maybe it doesn't matter but it does raise some questions about the relationship between the perception of time and the passage of clock time; between perception and object of perception and between experiential accounts and technical measurement of time.

DOI: 10.1057/9781137275363

Bolt does not go as fast as usual, but neither do his competitors. Perception of time is relative. Bolt is fast enough to win, but it is a 'slow' race with speeds of over 10 seconds, well in excess of the best speed of Usain Bolt. Bolt is slowing down, saving his energy for the final; time is energy as well as being calculated as distance divided by speed. Although the speed of each athlete is carefully and precisely recorded, time depends on tactics and context as well as being a simple formula. Even the men's 100m, the race which takes so short a time and whose participants vie for the title of the 'fastest man on the planet', is made up of segments of time.

Similarly, the women's race is subject to the same temporal divisions, into three segments as the men's, but lacks the enormous kudos that remains attached to the winner of the men's race; winning the men's race earns one the title of being the 'fastest man on earth' – a claim in which man is used to elide with human. It is now possible to be the 'fastest woman on earth' since women are now permitted to compete in the 100m event; a title won by Jamaican athlete Shelly-Ann Fraser-Pryce in 10.75 seconds at the London Olympics.

The 100m is not only the fastest competition in athletics but is also a sporting event which demonstrates, most powerfully, the mix of time, motion, flesh and politics. The 100m shows how time is social and political in diverse ways, each of which includes bodies in motion. Movement is highlighted by the speed of the competition. More than any other event the 100m is used to celebrate the Olympic motto of Citius Altius Fortius – faster, higher, stronger – and with a focus on speed, and its competitors, especially, at least hitherto, its male competitors, the event has achieved an iconic status in which temporality is inextricably embedded.

Because of the race's celebrity status in creating human beings who can be described as having the capacity to move faster than any other, it generates both interest and controversy. Speed is highly valued in this context and in the sporting history of track and field events and also in the wider arena of human achievement as defying the constraints of time. Debates have raged about which 100m athlete deserves the highest accolades. The 100m is not only the fastest athletic event, the race in which ever-faster speeds are recorded, but is also the event which has attracted the most social, cultural and political attention. The 100m is a good example of the social and cultural forces that are at play in the moment of spectatorship as the above experience of 2012 demonstrates. The excitement of the process of watching and of being caught up in the

DOI: 10.1057/9781137275363

event as a viewer is greatly enhanced by being able to situate the enfleshed practices of the athletic activity in its histories and aspirations, some of which are embedded in the timed records which highlight this particular event's capacity for speed. In part, it is the knowledge of the details of past records and understanding the strategies that can be deployed to improve chances of victory in achieving ever-faster times that makes each occasion and each competition more exciting. There is more to the 100m than these social-temporal elements which link the past and the future to the present. The 100m also has a political legacy which is a testament to the powerful forces of inequality which have traditionally pervaded sport, not only reflecting those which underpin social relations in the wider field of social and cultural life but also constituting racialised infrastructures and networks and those informed by inequalities based upon sex gender, class and disability.

The associations of being 'the fastest man on earth' with traditional masculinity also leads to the legendary and heroic connotations of the race. Men's sport has particularly powerfully developed connections to heroism based on the triumph-over-tragedy narratives of using sport as a route out of exploitation and disadvantage, as are particularly prevalent in men's boxing, for example (Woodward, 2006). Whilst men's boxing manifests strong associations with the politics of race, so does athletics and especially the 100m. The higher the status of the competition, the more investment is made in those who succeed. The visibility of the Olympics makes the games a very public platform for the celebration of the host nation and of each of the participating nations. The 100m is a particularly important platform because of the speed of the race and its possibilities for laying claim to the triumph of human beings over the limitations of time. The 100m matters socially and politically as well as being a source of gripping entertainment.

Pellon McDaniels claims that 'in one 10 second moment Jesse Owens overthrew white supremacy and black inferiority' (Olympic Stories 100m, 2012) when he defeated Hitler's expected champion, Eric Horkmeyer (who came 5th) and other white athletes at the 1936 Olympics. This is more than the reiterated sporting hyperbole as the 100m, of all track and field events is the one that has particular iconic and heroic inflections, is notably conflated with heroic masculinity. This major political event brings past and future into the present in particularly dramatic ways. What would happen if Owens won and the future possibilities if he did are all part of the present of watching the race itself? It is almost

DOI: 10.1057/9781137275363

impossible to isolate the different elements of the event itself; it is exciting and competitive but in such a situation it is impossible to disentangle politics; cultural and social practices and inequalities; and the enfleshed activity of the fastest race in sport.

Before the 2012 games, much of the discussion about the 100m and what it could deliver and what possibilities the race afforded centred on the relative merits of whether Carl Lewis or Bob Hayes was the greatest ever 100m runner. Such discussions encompasses bodies in time, the tactics of timing and the social and cultural making of time; this is also how legends are made. Such debates were the subject of many of the media discussions which provided both information and necessary background knowledge, especially of the warm-up activities in the period prior to the games. Heroic narratives, sporting stories and the making of legends combine with well-argued debates in which considerable expert knowledge is enlisted to support claims to promote more and more enthusiasm for the mega event par excellence of the Olympics. The more debate and the more knowledge circulates, the more the cultural centrality of the sporting event and the more the excited fans and followers are likely to be about the significance of the event. The BBC television series *Faster Higher Stronger Olympic Stories* which started on Monday, 9 July 2012, from 7 to 8pm BST took its title from the Olympic motto series of four programmes in the run up to the games (Olympic Stories, 2012). Starting with the 100m is likely to have been a way of kick-starting interest and acknowledging that speed counts, especially on sport and that superlatives are commonly attributed to success in operating at speed. The men's race is faster than the women's at present but one wonders if women were ever to exceed men's speed, rather than just catch up with previous records as they have in most sports, would women then be the fastest human being on the planet as men are now, mostly the fastest man on the planet? To be the fastest man assumes that you are the fastest human being whereas to be the fastest woman is just that, the fastest woman but not faster than men. Speed is culturally and politically valued along gendered and racialised lines.

As is argued in Einstein's theory of relativity, 'time is relative to those who observe and the frames of reference which they use, of course there is no such thing as "now" independent of a system of reference' (Barnett, 1957: 41). This is part of my frame of reference, but it is acknowledged as relative and constituted through my critical judgement of the sporting event and its translation and interpretation through the systems to

DOI: 10.1057/9781137275363

which sport is subject. The body practices and regulatory frameworks, including the mediatised processes of representation and systems of governance of sport, are constitutive of its times scales. This is strikingly apparent in the 100m when it is impossible not only to note what I experience as a spectator, but also, because of the speed of the race, for very different reasons, which places the event in its history, however fast it is moving now, to disentangle experience and sensation of the race now from previous records set and from other athletes' achievements and hopes for the future. I approached the event wondering if records will be broken as well as who is going to win and try to note down the feelings as fast as I can. Which strategy will triumph is more a matter for the post-race consideration.

'Real time': feelings and categories

Sensation may be direct and unmediated in the moment (Deleuze, 2005) but it is nonetheless social and political. Exploring time in sport means disentangling some of these social dimensions as well as the multiple forces which make up the event, but the sporting event, as my 100m example shows, is immediate and generative of emotion and affect which condense time and motion into the 'real-time' moment. Time is one of the elements along with other material forces. Time, however, has more immutable properties and its passing is inevitable.

For the researcher some of these processes are more powerful and forceful and influential than others. Time is understood and deployed in different ways in different sports. For example in cycling, the race is called a time trial because time is central to it; the time trial event demands knowledge of spectators beyond spotting your favourite and gauging where they are in the competition.

In the 44 km men's cycling time trial on 1 August 2012, the competitors set-off at 90 second intervals. The race is measured spatially by kilometres; the winner is the competitor who achieves the fastest time, but for the spectator it is not immediately or easily apparent who is winning or how the placements play out. The event is further complicated for the researcher and the fan by the presence of Bradley Wiggins, winner of the 2012 Tour de France. Wiggins has the distinction of being the only Englishman ever to win the competition. Wiggins represents the hopes and dreams of British fans but his presence raises questions about the

DOI: 10.1057/9781137275363

connections between the past and its influence on the present as well as hopes for the outcome of the race. Will he be too exhausted to win? No, he can win. Wiggins stops the clock at 50 minutes 39 seconds, 42 seconds faster than Tony Martin the German champion and 68 seconds faster than his British team mate Christopher Froome.

The date is 30 August 2012; at 10.03am BST it is the 400m swimming S12 qualifying heats. The crowds in the aquatic centre cheer as the swimmers enter the water. There is enormous support for GB's Hannah Russell, an S12 aged 16 for whom expectations are high. S12s have very limited vision; they can see no further than 1m ahead and recognise the shape of a hand. Hannah Russell wins easily even if she doesn't quite break the world record. There is a huge gap between fastest and slowest competitors, between Hannah and the last two to come in; nonetheless there is a great cheer when the final competitor touches the end. Paralympic swimming looks much the same as Olympic swimming but there are complex classifications of visual impairment ranging from some mild restricted vision to being completely blind, where athletes are permitted to have a tapper to provide guidance. In swimming that combines the four Paralympic categories of limb loss, cerebral palsy, spinal injury and other cross-disabilities (Paralympics Classification, 2012: 50). It is very complicated, but as Hannah says, when interviewed, her event is about the ability to compete within a particular time (and in her case to go faster than anyone else in her heat).

These classificatory systems which categorise degrees of disability within which athletes can demonstrate their timed abilities are all part of the sporting event, including competitors, spectators and commentators. Through the experience of spectatorship and of immersion in the field, in particular in the knowledge generated by and implicated in sporting practices, including the rules of the game, it is possible to gain greater understanding of tactics and timing; of the measurement of time and of how it is lived and used by athletes.

The point about an emphasis upon the local system of 'now' is that it prioritises the relationship between the enfleshed practices of the sporting event, the spectator and the media construction of immediacy. Commentary in the moment is distinctive in sport and is constitutive of the event itself, produced within particular discursive regimes. Even when present at the event, for example at the ExCel watching boxing, the intensity of spectatorship is made up of expressions, affects, sensate experience, language and the social and cultural context of the

DOI: 10.1057/9781137275363

experience of 'now', which has particular resonance in boxing (Oates, 1987, Woodward, 2012b).

Time and motion

Time and motion are components of temporality that cannot be disentangled. Sporting times are moving times; bodies in motion and timed movements. Sporting bodies are in motion in the field, track, pool and ring and sporting times are made through change in the to and fro of making histories. Body time (Adam, 2004:101) does not only mean birth, death and the corporeal experiences of ageing in between; body time is also motion, strength, speed, which are the *citius, altius, fortius* of the Olympic ideals. Some of the pleasures to which flesh is heir to include experiencing the body in time, with time and against time.

Time involves movement from A to B, from start to finish, from then to now, with the promise of future becoming beyond now and each of this movement has particularly a strong force in sport when events are framed by completion and defined by the move from start to finish. The enfleshed activity is terminated after a set period of time marking the passage from start to finish and records and scores are fixed at this point. The research process too engages and is, in time, with its own movements between starting and finishing. It is not a linear process, rather one which involves iterative and dialogic movement, especially in recovering the past in relation to the present and formulating new possibilities and targets for the future, but there are at least some finishing points, for example when the project is completed and when the publications are delivered and there is always some initial point of engagement.

A focus upon time and times means looking at movement: the movements of bodies, which are of course central to sport (Woodward, 2009, 2012a), and chronology as well as the relationship between present, past and future, for example as expressed in the history of the Olympics (Guttmann, 2002) the relationship of the 2012 games to previous Olympics, especially Beijing in 2008 as the most recent, and looking forward to Rio in 2016. Understanding and recording time has evolved historically as 'clock time' (Kern, 1983) and in sport, as in most fields of activity and endeavour in the modern world, time is money. Within globalised capitalist economies, time becomes a commodity which like profit and resource is unequally distributed depending on the relationship with the means of production (Marx, 1973, 1976). As Max Weber demonstrated

DOI: 10.1057/9781137275363

in *The Protestant Ethic and the Rise of Capitalism*, time-keeping shifted to time accounting (Weber, 1989 [1905]) and the generation of profit became an end in itself. Time becomes a regulatory mechanism, which time clearly is in sport, although not always within so explicit a teleological trajectory. Movement in sport is not solely governed by the profit motive and clock time in sport is also evident in the pursuit of a personal best and even pleasure. Sport still has strong associations with play (Galeano and Fried, 1998, Guttmann, 2005).

Sport is concerned with all sorts of movement which are linked to change, which presents problems and challenges for the researcher, especially in terms of approaches to investigating what is moving and changing and the interrelationships between time and motion. These connections between time and motion are highlighted by the intensities of sport, whether as a participant or a spectator.

Sport is also located in the now, especially in what is experienced and seen and then recorded. Live broadcasts can take precedence and be more valued. For example, broadcasters, advertisements for the public big screens and pubs, along with other venues for group viewing, always exhort us to 'see the game live'. Whatever the iconic status of the sporting moments which are put into the discourse of history and constitute heroic moments such as the 1966 men's football world cup and Diego Maradona's goal of the century in the quarter finals in 1986; the 'Rumble in the Jungle' between Ali and Foreman in 1974; Jesse Owens at the 1936 Berlin Olympics, John Carlos and Tommie Smith on the podium at Mexico City and in the twenty-first century Michael Phelps's and Usain Bolt's gold medals; the 2012 games to mention but a tiny few such moments, 'live time' is not only news but has power to recruit interest and imagination now. In the present of 2012, Phelps's and Bolt's achievements, for example, invoke past moments and triumphs which heighten the significance of live time. Sporting times include the media coverage of sport, engages with the now, as well as discussing the now in relation to past performance and predicting future form.

Sports fans always demonstrate the strength of their attachments by laying claim to memories of great sporting moments, most notably of 'when I was there' (present when the event was live) or 'I remember where I was when I heard the score/found out the result/learned that the record had been broken' or just heard that it had happened (Boddy, 2008). The 2012 Olympics and Paralympics have produced many opportunities for such moments in the future.

DOI: 10.1057/9781137275363

The discussion in this chapter raises these questions about the points of connection between past, present and future in relation to motion and movement and about what a study of sporting times can offer in understanding some of these relationships and of the concept of temporality. Temporality is implicated in the links between time and motion which are explored by looking at the methodological implications of a study of sporting times, which include the approach taken in the book, which in this particular case involves writing during the 2012 Olympics. As we are aware, in sport and in sports media coverage and commentary showing an event as it happens in 'real time' is live and not recorded.

People in time: writing in time

The methods adopted in writing this book include writing as I watch as well, of course, as reflecting on the process and on the events in which I am entangled by virtue of both viewing the sport and writing about it. Writing in time cannot be simultaneous, but the process has particular resonance in sport where distinctions are made between a live and a recorded event. The short extracts written in 'real time' as I watched are reproduced more or less as I wrote them and thus include sometimes eclectic choice of what to say and what to exclude and a mix of what hit me at the time and what seemed timely. The experience and the strategies I have adopted are used to explore some of the specificities of sport and its appeal. The appeal is immediate, but the legacy may be less substantial. Or maybe the legacy is all the more likely to be enduring because the appeal of now is so powerful. Rather than thinking about what is so attractive about sport in the entertainment it can provide, I seek to consider particularly how time is implicated in the making of these attractions which are both transcendent of the moment and embedded in it through the very material properties of the bodies which take part and the sense which are invoked. The process of writing in the duration of a series of sporting events like the Olympics also highlights the ways in which time is so central to the constitution of social and cultural processes and how past, present and future are condensed and intensified in the moment.

I have chosen to reproduce extracts of what I have written during the games for events which I have watched on television. The approach would not be feasible or manageable for live spectatorship and so the extracts selected are all from those competitions and some interviews

DOI: 10.1057/9781137275363

which I viewed at a distance. For the purposes of this book I have only engaged with the question of the relationship between 'being there' and the sensations, including the implication of all the senses in the process of spectatorship at an event, insofar as these matters relate to time and temporality. There are intensities in the experience of being in time at a competition which combine with other elements in the assemblage of forces and make-up the whole event (Woodward, 2012b), but it is these issues of being in 'real time' which I wish to focus upon here.

My starting points were thinking about the centrality of time in sport with its stress on timed records and achievements and histories, memories and enduring legends along with the absence of critical analyses of temporality in most of the literature – either on sport or temporalities – and the problem of time of times in sport. Rather than being addressed as separate and discrete strands in the make-up of sport I seek to condense these different dimensions into the experience of 'real time' and of now which clearly, or perhaps confusingly, has affective, emotive qualities and strong elements of psychic investment and strong attachments, whatever the scientific and quantitative elements of times and timing in sport.

Sport seeks certainties and sets boundaries and rules which are enforced through sport's regulatory bodies (Woodward, 2009) and the parameters it sets and time is central to these regulatory frameworks; yet time its measurement and discursive powers seem to be assumed and to pass largely un-interrogated.

Being in time is often manifest in sport not only through the valuing of live over recorded time but also through the particularities of the moment. One example among many during the London games was the Men's Canadian singles canoe slalom on the 31 of July; commentators reaffirmed that this was an 'on the day event'. This is a comment made of many events and somehow suggests that there is something instinctual about the immediacy of the moment that no amount of preparation and training can address. On the day something particular, important and possibly unexpected happens. This is what informs my approach to noting my reactions and some of the commentary as it happens. This cannot be a direct experience without being located within the culture and context of the activity, not least because the event on television includes camera angles and visualisation and sound techniques and the commentary and interpretation of the production company; the BBC for the Olympics and Channel 4 for the Paralympics. However, it is time and its duration which in this instance provides the nearest possible route

DOI: 10.1057/9781137275363

into capturing the moment as I hope to demonstrate by a focus upon 'real time', however troubling the concept may be.

This raises questions about the relationship between consciousness and experience as well as perception and the objects of perception. The issue of consciousness is troubling in some ways in sport and has contradictory implications. Consciousness is important, specially in sport, because athletes often invoke the notion of timelessness and claim that they are not conscious of time when they are competing. Jim Hines, the 100m gold medallist who first broke the 10 second barrier, at altitude, in Mexico City in 1968, acknowledges that the race is quick for onlookers at under-10 seconds in the games but as a competitor he had no idea of time; it was infinite and absent; it was just experience (Olympic Stories, 100m 2012). Lack of consciousness of time and of time passing is more than the practical exigencies of running the 100m; it is also a transcendence of the routine manifestations of consciousness. The idea of timelessness is difficult to capture. Consequently my approach of writing in 'real time' is an attempt to access something of these sensate experiences and sensations when time is either transcended or reduced to an irrelevance.

Barbara Adam says time is human and one distinguishing feature of humans could be their consciousness of time through memory and awareness of future time and the duration of time. Adam uses examples such as human cognisance and understanding of how long a pain or a journey will last to illustrate human capacities which animals largely lack (Adam, 2004). However, these human properties also include the possibility of acknowledging timelessness and moments, somewhat surprisingly during a race in which success is entirely measured by timing, which also characterise experience. The researcher too can be lost in time, and the suspension of being conscious of time is one aspect of endeavouring to capture such impressive and exciting corporeal achievements by writing in 'real time' which is part of the process. In endeavouring to write in 'real time', during the games and during actual events, lays the spectator and critical observer open to being caught up in the thrill of the experience which can render the viewer temporally incapable of anything but enjoyment and make critical analysis impossible, albeit only for the duration of the event.

There are other routes into immersion and participation in the field of research. Sports studies and research in sport, especially ethnography and practices of participant observation, have generated discussion about the relationship between the researcher and the field of research (Cashmore, 2005, Horne et al., 2012, Wacquant, 2004, Wheaton, 2004, Woodward, 2008)

DOI: 10.1057/9781137275363

which is located in the time of lived experience. Being in time can be evident through the embodied presence of the researcher as well as through the identifications of spectatorship and virtual participation (Sugden and Tomlinson, 2011). In sport research there are powerful links between participation as a supporter and the process of conducting research which involves being there, such as sharing lifelong commitment to a sports club, for example, in football (Giulianotti, 1999, Giulianotti and Williams, 1994, Goldblatt, 2007, Wagg, 2004). The sporting attachments which are often strongly connected to sex gender (Wheaton, 2004, Woodward, 2006, 2012a, 2012b) can be subject to critical consciousness even when writing in 'real time' and as a researcher I am aware of the forces in play in what is covered and how events are presented and represented.

Such representations are more part of the reporting of events, post-hoc than in the live time of the competition. This is particularly evident in 2012 and is another justification of my methodology in seeking to link real-time observation to critical analysis and reflection. This approach permits an exploration of what is 'real time' and the inextricable connections between social forces, past achievements (and failures and struggles as well as what dreams and hopes are brought to the event) and future aspirations to present performance in the study of sporting times and of temporality more broadly.

Conclusion

Time, especially as experienced in the now of the Olympics seems at one level to be concerned with the material dimensions of the physics and the science of the fitter, faster, stronger body, but it is also social in a variety of different ways. Some bodies and their timed achievements of particular events are valued more highly than others. Some times are called into question when it is the wrong body which has achieved them, as in the case of Ye Shiwen, the Chinese swimmer. It is not only a matter of how fast the body is propelled through time and what is the recorded speed that generate the record or assures victory, qualification or even a medal. In the case of the Chinese swimmer Ye Shiwen, in 2012, who exceeded the record breaking speed of the US male swimmer, medal winner Ryan Lochte for the last 50m of her race, the experience of watching (when the viewer just knew that she was going pretty fast, but could make no gendered comparisons) in 'real time' was only that this was a

DOI: 10.1057/9781137275363

major achievement for so young an athlete. In this instance the protest came later in relation to the protests and accusations of malpractice, all of which contribute to the social implications of time. In this case the social contextualising and the knowledge came after the event. At other events social forces are embedded in time and are constitutive of the experience of the timed sporting event, as must have been the case at the 1936 Olympics when Jesse Owens triumphed. Time changes things. In 2012 the high achievement of black athletes becomes ordinary in that it is taken for granted; becoming beyond race in some aspects. The matter of bodies is always social too and thus measurement does not occupy the realm of physics untainted by the social. Some bodies are permitted more time than others and distances are measured according to competitors. The 100m is still largely a men's race, especially in the attention and hyperbole which the men's competition accrues.

The 100m is an event which is more intense, because of the speed in which it is run and its very short duration. The athletes may have time to work out their tactics, Usain Bolt may have time to look around, but the rest of us barely have time to blink. The 100m creates unmediated sensation which seems divorced from its social context, which is not because social, economic and cultural forces are all caught up in the event, however personal the experience of time may seem, even timeless in the intensity of the moment of now.

The methodological framework of this book draws upon the implications of psychosocial notions of reflexivity, especially in bringing together the social and cultural discursive contexts in which sporting times are lived. There is also an ethnography of spectatorship which uses approaches developed in anthropology and feminist participation, driven by a recognition of the importance of social forces (Woodward, 2011), especially those through which inequalities are forged (Woodward, 2011) and the need to question fixities and explore the possibilities of change. The trope of writing in the present and of focusing upon the present is one method adopted, albeit with the acknowledgment that the present is always connected to the past and the future. Unlike many of the books which have engaged with temporality, however, not only is my focus sporting time but also I start with the experience rather than the theory.

I use short extracts of what I wrote during the games, while I was watching, although it was not always possible or practical to complete the writing, however brief within the same period of time as the event, of course. My notes are brief because of the pressure of time in a process

DOI: 10.1057/9781137275363

which puts the observer and commentator under constraints that are in-part in common with the athletes (but without the same enfleshed degree of preparation). Time is not the only consideration however. Being caught up in the event, in the to and fro of affect between spectator and athletic performance and all the elements of light and movement which make-up the sporting activity, is not a time for reflection but rather one of immersion in the moment and in the sensation of the activity (Woodward, 2012b). The selection, which is made of what to note, is also immediate and not crafted in any way. However, as the example of the 100m in this chapter demonstrates, this process of writing in 'real time' affords some insights into how past, present and future are condensed into the 'real time' of now and it is the legacy of the past and the projections into the future informed by past knowledge and present sensation which makes the experience meaningful and comprehensible.

The focus upon movement which a study of time and temporality demands is combined with methodologies which prioritise the immediacy of the sporting moment and the much vaunted sporting celebration of the live event. The live event is itself problematic and often not as live as it claims to be, as I shall demonstrate. But by being attentive to the live event and by attempting to capture what is happening in 'real time' does raise some interesting issues about how time works and how time is experienced as well as how it is manipulated. There are mobilities which do not follow a clear chronological and linear path; the past, present and future are entangled and social forces operate to make the connection between temporal, social, cultural and political factors closely interconnected.

One way in which past, present and future are bound together is through the making and remembering of memories, which are strongly connected to the past and to the idea of summoning up in the present what has happened or has been felt at an earlier time in the making of legends and myths, which is so much a part of the culture of sport and sporting times. Time in sport is not only concerned with record keeping and record setting, it is also about making and re-making the stories through which the social and cultural implications and meanings of sport are transmitted. Sporting times are concerned with a whole range of time-based records which then become embedded in the stories which make up sporting times and sporting histories and legends. Memories of times achieved and beaten in sport become its myths and legends and make its future.

DOI: 10.1057/9781137275363

3
Memories

Abstract: *Chapter 3 explores the to and fro of remembering and re-remembering which is frequently invoked in sport, whether in commentary, following and fandom or in the enfleshed experiences of training which is so often informed by the observation and rehearsal of previous performances. In the Games, past records are re-visited and re-invoked in the run-up and in the coverage of the events. This chapter looks at what is and what is not remembered and how these memories are made and remade, perceived and expressed. Memories are reproduced through the narratives which pervade sport in all media but this chapter uses Bergsonian understandings of the movement of memory to challenge the linear chronology of sports storytelling and to look at who is silent in the stories and at who speaks and is spoken. Sporting memories are profoundly connected to social exclusion, notably in terms of sex gender, which has presented an area often marked by silences, invisibility and absences of women in sport, but which is also part of changing times; the excitement and exhilaration of women's events from rowing to boxing in 2012 showed the promise of changing times so well.*

Woodward, Kath. *Sporting Times*. Basingstoke: Palgrave Macmillan, 2013. DOI: 10.1057/9781137275363.

 DOI: 10.1057/9781137275363

Memories matter enormously in sport. Remembering can take many forms, ranging from the capacity to recall scores, results and final times – notably either of records broken or of records lost – to the stories and legends through which we make sense of the field and of its meanings and, especially, its powerful attachments. We need heroic narratives and stories of endeavour and of triumph-over-tragedy in order to come to terms with the place of sport in the contemporary world. At the time of writing in September 2012, it is possible to see legends being made through mediated selection processes, for example in the emergence of commemorative collections of 'London 2012'. Heroes of the past are made in the present as are some of those of the future (*Independent on Sunday*, 2012). Social forces are in play too in the making of memories. These are all part of keeping the project of legacy going and maintaining the intensities of 'real time' and of being there even if virtually. In sporting mega events, like the Games, being there does not always require one's corporeal presence in the stadium, and feelings of collective belonging can be sustained even at the big screen viewings in public places. Watching at home still provides the knowledge through which connections can be made, even with complete strangers, at a time when it seems everyone is talking about the games.

Sporting stories also offer entertainment and provide mechanisms for linking the embodied practices of sport to social, cultural and political forces. In the Olympics, one of the most visible presentations of what constitutes narratives of the games and the host nation is the opening ceremony. Opening and closing ceremonies present opportunities for engagement with memory practices and the making of memories through showcasing the event and the nation by constructing narratives in which the past in made material and the future is signposted and envisioned.

London 2012 started with a distinctive opening ceremony which had been preceded by memories of Beijing in 2008 and discussion of the possibilities. What could London do to replicate or rival even very modestly the spectacular displays of Beijing? The small-scale handover at the closing ceremony in Beijing suggested much more limited ceremonies in 2012. London 2012 started well before 2012, of course, and well before the handover in Beijing in 2008 (Beijing Closing Ceremony 2008, 2012); it started with the hopes and dreams which became the bid. The eight-minute segment in which London organisers presented a very modest taste of what London might offer was a tiny taste which avowedly could not compete with the splendours of Beijing. A local mix of a red double-

DOI: 10.1057/9781137275363

decker bus, gold medal cyclist Chris Hoy (subsequently Sir Hoy) dressed as a city gent accompanied by fellow British cyclists reflecting team GB's success in cycling, popular singer Leona Lewis who is the winner of a television talent show. The X-Factor, and culminating in megastar footballer David Beckham, kicking a football into the crowd, before the action turned to a 'memory tower' in the centre of the stadium acknowledging the centrality of making history and making memories through the mechanism of the Olympics. The Olympics 2012 might even have started as the legacy of 1948, the last time the games were hosted by London in the aftermath of the privations of the Second World War. The Olympics of 1948 was the austerity Olympics, a label which sums up the social implications of the sporting event and the social and cultural implications of time. It was both the time of ration books held in a London still war-ravaged with extensive bomb damage, and of hope and expectation of peace and, especially in relation to the Olympics, of the possibilities of materialising the principles and democratic dreams of the Olympic Movement (Hampton, 2008) not least because the 1948 Games were the first Olympics after the political hostilities of the 1936 Berlin Olympics – so-called Hitler's games.

The London games in 1948 provide a particular example of the ways in which memories are made and which elements of memory matter and become material in the now of 2012 amidst the enthusiasm and collective engagement with the games; earlier times are invoked and revisited. 1948 is also part of a process of recall which highlights the 'people's games' because of the ways in which, in the circumstances of economic crisis which followed the Second World War, despite of lack of resources (certainly nowhere near the massive funding which more recently has been invested in sport) the games were a great success.

Sport is regularly the site of listings of the greatest contestants, the most exciting moments and debates about a mix of personal and collective memories and the recorded, measured criteria of success. However, individuals are likely to draw upon personal lived experience within their own biographies, collective memories, and certainly the lists that are compiled (for example BBC Great Olympic Moments, 2012) through classificatory systems which combine record achievements, such as those of Usain Bolt in 2008, or Carl Lewis in 1984, a first for a category of person such as those generally under-represented in the annals of great Olympians; transformations of sporting techniques, such as Olga Korbut's radical redefinition of gymnastics; heroic endeavour often combined with

DOI: 10.1057/9781137275363

personal achievement, such as the performance of Emile Zatopek in 1952; or an achievement that might not have been expected, like Chris Hoy's in 2008 or the GB men's hockey team's in Seoul in 1988; and sometimes triumph over tragedy, such as the sprinter Wilma Rudolph in Rome in 1960. Memories are not always celebratory; Ben Johnson's performance may have been stunning in Seoul in 1988 but the memories are tainted by his subsequent positive testing for steroids. The Olympics have provided a site for political interventions, including the expression of protest (Lenskyj, 2007, 2008). Mexico City was the site of what must be one of the most memorable challenges to the infrastructures of social exclusion in the explicit rejection of racism in the black power salutes of John Carlos and Tommie Smith, as well as also being the site of tragedy, as was expressed in the deaths of Israeli athletes in Munich in 1972. All of these events are legitimised within the processes of memory-making which is a collective and, especially, social and cultural (and political) project rather than introspective recollection of lost times. Even personal introspective accounts, of course, are social not least Proust's own reveries and reflections on the perception of time and its effects (Proust, 1982).

Memories as formulated through either great moments or stunning and spectacular times which might have fame because they have notoriety involve not only a dialogue between past and present framed by future aspirations but also powerful social, political and cultural forces which are always complicit with what is included and what is excluded. Disruptions may be triggered by what appear to be individual actions, such as those of John Carlos and Tommie Smith in Mexico City in 1968, but they are located within a wider social and cultural terrain; the disruption relates to whose voice is being heard and how interventions can unsettle the existing social and political order. In the to and fro of memory-making the idea is perceived before it is represented (Bergson, 1896 [1991]) and there is a delay in the making of memorable moments, although in 2012 there has already been an explosion of such claims. However, the passage of time is necessary for the memories to become stabilised. Memories cannot be put into discourse now but have to be part of a process of iteration and remembering and re-remembering over time.

This chapter explores the to and fro of remembering and re-remembering which is so frequently invoked in sport, whether in commentary, following and fandom or in the enfleshed experiences of training which is so often informed by the observation of previous performances. In the Games, past records are re-visited and re-invoked in the run-up and in

DOI: 10.1057/9781137275363

the coverage of the events. This chapter looks at what is and what is not remembered and how these memories are made and remade, perceived and expressed. Memories are reproduced through the narratives which pervade sport in all media but this chapter uses Bergsonian understandings of the movement of memory (Bergson, 1896 [1991]) to challenge the linear chronology of sports storytelling and to look at who is silent in the stories and at who speaks and is spoken. Sporting memories are profoundly connected to social exclusion, notably in terms of those of sex gender (Woodward, 2006, 2012a, 2009). Thus performances which go down in the annals of sport, rather like other public myths, legends and heroic narratives, tend to focus upon the activities of men and masculinity rather than on women and femininity; sex gender has been coded in different ways (Woodward, 2012a). There are possibilities of change and 2012 opens up opportunities for the transformation of memory in that there is new material through women's participation in the games which makes new memories possible.

Memories and the idea of looking back on past Olympics and Paralympics as well as celebrating the present and enhancing the experience of the present through its location within a chronology which draws upon the past and its achievements are central to the organisation and structure of the games and given expression in the rituals of the opening and closing ceremonies. The configuration and expression of memories move between opening and closing ceremonies in the now; the histories and heritage of the host nations are embedded in and celebrated by the opening ceremonies, none more so than in 2012, which presented somewhat idiosyncratic and critical set of narratives of Britishness. The closing ceremony looks forward to legacy and establishes its own narratives of Olympic continuity as well as some celebration of the games that have just been enjoyed.

Legacy – opening and closing ceremonies

Friday, 27 July 2012. This is the start of the 30th Olympiad. It starts with a swift montage of the British Isles: Isles of Wonder. The eye moves over an eclectic display from cricket greens to the London Eye accompanied by the Clash's London Calling and a brief extract of John Lydon's irreverent version of God Save the Queen. This looks like something to challenge the linear narrative. The diversity of the combination of elements makes

DOI: 10.1057/9781137275363

this a very innovative assemblage. Balloons burst and Tour de France winner Bradley Wiggins takes to the stage in his yellow jersey. Then Jerusalem with all its resonance of England's (in an elision of Englishness with Britishness) green and pleasant land sung by a lone choirboy, followed by Danny Boy for the Irish.

The mix seems designed to challenge as well as instate a story of Britishness, or maybe Britishness is so complicated that it cannot be made into a conventional narrative. Anyway this looks like a lively take and even a subversive one. Rather like celebrity spotting I find myself trying to pick out things that fit. I am not sure what they would fit into, but it is probably a remembered version of what the story of the British is as taught in the history books of imperialist rhetoric. The massive bell is another clue to Britain's industrial past.

The actor Kenneth Branagh is recognisable although I am not too sure about the part he is playing. After some time I realise that he is Isambard Kingdom Brunel, after making links with the smokestacks that have emerged from the ground, along with factory workers and the marching suffragettes. The workers look very happy and pleased with their lot which suggests little engagement with a critical Marxist analysis of class such as Friedrich Engels' *The Condition of the Working Class in England* (Engels, [1845] 2009), but this is a party and clearly a celebration of Britishness. The Industrial Revolution is clearly identifiable as a strand in the British story – what made Britain great along with significant visibility given to women's political struggles – and I also see later that the CND symbol is one of the formations. At a time when Britain is fighting in Afghanistan and is committed to nuclear technology, a reminder of the politics of nuclear disarmament is clearly important – or probably not as visibly clear as you need to be observant to catch it. Isambard Brunel reading Caliban from Shakespeare's *Tempest* is a bit puzzling except that industrial progress and innovation and Shakespeare are part of Britishness.

The Olympic rings fill the stadium and the sky with light in a more conventional deployment of light imagery but the most subversive and unpredictable plot of the artistic director of the 2012 games in Danny Boyle story of Britishness is what comes next. James Bond, lent authenticity by being in the person of by Daniel Craig in evening dress appears, not at the stadium but apparently filmed at Buckingham Palace, clearly intent on an audience with HRH Queen Elizabeth II by parachute. She seems to be expecting him. Maybe he's come to take her to open the games? They leave together and the next thing they are flying in a

DOI: 10.1057/9781137275363

helicopter from which the Queen leaps to arrive unruffled at the opening ceremony. This is mind-blowing stuff as the Queen is clearly party to this. It did look as if the stunt man sown into a dress so similar to the queen's (as we later discover) really is the Queen. This is both a timely illusion and one which is choreographed to achieve maximum impact. Not only are the sporting events carefully timed!

The participants are not all famous; indeed this display seems to rely heavily on volunteers and the British public including a large number of NHS employees in the mix of great past moments, strands in the narrative and more recent strengths of the British way of life and culture. The jolly NHS nurses are reminiscent of Denis Potter's fantasies of singing medical staff in his play *The Singing Detective*. This is not as surreal as the octogenarian Queen, the dignified element of the English Constitution in the words of Walter Bagehot (Bagehot, 1966) demonstrating greater bravery and courage than James Bond by parachuting out of a helicopter. This is an immediate response and not a considered reflection, but there is always a relationship between perception and representation, and unmediated sensation (Deleuze,1989) is not un-located; perception too is contextualised by what has gone before.

Memories are remade and selected in this idiosyncratic display which both reinstates and subverts conventional storylines. Stories and the making and re-making of narratives are a crucial component not only of making sense of the world but also of making memories (Ricoeur, 2006). It is not only the linear narrative which is challenged. What appears is at times the muddled personal memories we have upon which it is difficult to set a clear structure or to put into a coherent pattern. Danny Boyle seems to be able to reproduce these personal confusions of which we endeavour to make sense. Boyle's intentionality and the processes through which decisions are made about so large scale an event are part of the process of spectatorship. I am not able to watch without thinking why these images, formations, activities have been selected. It is hard to become immersed in 'real time' for the ceremonies because of the conscious processes that are part of the assemblage which makes memories through presenting stories – however eclectic and mobile those stories are in this case. There appears to be some awareness that authorial control is not paramount and whilst light and scale are deployed to manipulate response and plug into particular memories, there is some acknowledgement of the decentred author. In this assemblage of feelings, light, sound, people and things, the emphasis on the unexpected and the idiosyncratic through its take

DOI: 10.1057/9781137275363

on Britishness as innovative and eccentric generates the possibilities of creative responses with some understanding that texts can be received as well as presented in different ways.

They are also humorous stories which subvert authority, as is the case with the parachuting Queen and James Bond. The contradictions become visible and comedy is used to mock some of the sentimentalised fixity of certain cultural memories. The Chariots of Fire sequence is an example of this. Simon Rattle appears to conduct the London Symphony Orchestra. I have this information because I already know who he is. This is reassuringly predictable and an expected narrative invoking the familiar beach scene filmed at St Andrews in Scotland for the film *Chariots of Fire*. In my head I am already rehearsing the scene on the beach and stories of athletes Harold Abrahams and Eric Liddle coded by the tensions between social and cultural constraints and personal commitments. I do not expect the appearance of the comedy actor Rowan Atkinson in his role as the humourous figure Mr Bean, firstly playing at the keyboard, looking bored and irritated at the emotive cadences of the signature music before featuring in a comic reprise of the beach scene and tripping up his rivals.

The ceremony continues in compliance with IOC Charter demands, for example the procession of participating nations in alphabetical order and speeches by the LOCOG chair, Sebastian Coe, followed by that of the chair of the IOC, Jacques Rogge. The finale reverts to the spectacular and really is spectacular insofar as the Olympic flame ignites the cauldron which is to remain alight for the duration of the games: traditional practices but with artistic and specialised scope for innovation by the host nation. What is particularly interesting about this opening ceremony is not so much its use of technologies to purvey the spectacular but its mode of storytelling and its reformulation of memory, not only through the to and fro of past and present but its self-conscious disruption of a linear narrative either in chronology or through the rhetoric of Britishness. There is not one story but several in a celebration of discourses, like the stratified deposits of common sense (Gramsci, 1971), made up of political events and movements, economic processes, systems and practices, pivotal moments, literature, music folktales and cinema. Another distinctive feature of the 2012 Olympics opening ceremony is its celebration of ordinary people, transforming official historical accounts and public moments into a democratic party and moving between actualities, lived experience and cinematic representations, thus engaging with the realities of the everyday and public structural transformations.

DOI: 10.1057/9781137275363

Opening ceremonies are concerned with promise and the utilisation of the past to construct the promise of the future in the present. Whilst this is a distinguishing dimension of Olympic opening ceremonies, it is also instructive about the processes through which memory operates in its claims to the recovery of the past. Such events demonstrate well the to and fro of memory and its fluidity, which lies not only in the difficulty of ever reclaiming what has already passed

However complex the assemblage, nonetheless, it still affords a way of providing connections, albeit sometimes forged through ambiguous links and by encompassing disruptions and contradictions, which make a set of stories and, in this case a version of a narrative of Britishness, by recovering memories. These memories are made in the present and embrace possible futures as well as being explicitly and directly drawn from established histories – as in this case, from the legacy of the Industrial Revolution.

Memories inform perception and are explicitly deployed, albeit in ironic and humorous ways which are gently mocking of authority, in the opening ceremony to the London Games. These rituals are as much events as the competitions of the games, but there is a distinctive mix in which intentionality and agency of different actors are caught up in the mix, and London 2012 is marked by agency being afforded to an artistic director who may indeed be going with the times, but is stepping outside some of the expected traditions and flows of continuity, especially by introducing humour into the moment (BBC Olympics Opening Ceremony, 2012).

Looking back and moving on

The closing ceremony of the Olympics demonstrates the way the sporting practices of the games themselves construct short-term memories and involve some dialogue between opening and closing as well as summoning up some of the achievement of the games themselves through the recapture of pivotal moments. The discourse and rhetoric of the closing ceremony is also indicative of the ways in which what is put into discourse provides us with the matter of memory. We find out what we are likely to remember. In 2012 this applies more to the pre-ceremony build up than to the actual closing ceremony of the Olympics. At least this is what happens in the build-up to closing ceremony with the

DOI: 10.1057/9781137275363

BBC's review of the games' time and temporalities that are built into the rhetoric and hyperbole. I watch this programme on BBC television on 12 August 2012, in the early evening two hours before the 9pm start of the London Olympics closing ceremony.

The build-up reviews the achievements and immediately highlights the centrality of movement to the games. Time is central because it matters in sport, where timing is crucial; this event is located in Greenwich Park. Time and the measure of Greenwich Mean Time is the standard of these games. Timing techniques too immerse the viewer in the experience of being in time. The visuals are even speeded up to show movement of people through the Olympic Park then slow down for athletes crossing the line, such as Jess Ennis in the heptathlon final event, the 800m – and at this stage I know what happened. I am no longer waiting to see what the outcome will be but I am wondering which events will be picked. I pick Jessica Ennis because I know she will be mentioned, and I am a keen supporter of her as a spectacular athlete who comes from Sheffield.

The ceremony engages with the senses: not only is there contrasting visual representations and timings but there is the musical accompaniment which is quasi religious and militaristic, albeit informed by shades of Hollywood, that is used to lend dignity and grandiosity to a spectacle that I feel needs no such assistance. The memories of recent triumphs are so enjoyable nonetheless.

I see that selection does not depend only on level of achievement. Some of the recognition depends on when you achieve your success; for example David Rudisha who is the only man to run under 1 minute 41 seconds in the 800m received limited acknowledgement because it was the same day as Usain Bolt's medal. Bolt is man of the Games even though Michael Phelps has won more medals than anyone else.

We learn that the GB men's gymnastics team's first medal in 100 years, a bronze, came on the '5th day'; an expression resonant of biblical rhetoric. In order to invest in greatness, sport has to mimic the language of greatness; a discourse that has more currency in the affairs of state and the rituals of religion than what is still associated with play, somewhat ironically given the level of personal and collective attachment and commitment it generates. Nonetheless sporting heroes and legends are important in this context and are given the respect they warrant. 'Making history at Eton Dorney' (the location for the rowing and canoe sprint events, which was constructed and is privately owned by the independent school, Eton

DOI: 10.1057/9781137275363

College) is about the two women, Helen Glover and Heather Stanning, who won the first Team GB gold medal. 'What had happened slowly was gathering pace'. Bradley Wiggins comes on to celebrate his victory, made all the more greater by having already won the Tour de France, which is part of the present. Wiggins places himself in time referring to the number of medals he has won; 'what's most important is this is four so I have to plan for the future and number five and Rio'.

Speed applies not only to achievements in track, field, ring and pool. The day Mo got his gold there were '3 gold medals in 45 minutes. Who could follow that?' The response is 'more Brits'. Sporting times are patriotic times, but selectively.

These events are 'etched in your mind for ever', says Denise Lewis, 'for 40 years' – she goes on making the time span more real by setting a time limit. What we say now if experienced with sufficient intensity on the day it happened-now can be measured by claims forever or infinity or at least 40 years. When you say forever you mean it feels like forever now; the puzzle this poses concerns whether this is a rhetorical device or, as I want to argue something which is indicative of the relationship between the experience of time, the perception of time and attempts at measuring and quantifying time.

Memories are invoked and utilised in different ways, however, and the strategies and mechanisms that are in play tell us a great deal about what the making and understanding of memory in the field of sport demonstrates about how temporalities work. Memories inform the claims of the programme and especially lend authenticity to the status and credibility of sport as a serious topic, one which matters in the scheme of things. Sport is part of the myths and legends through which we make sense of being in time and of being human (Adam, 2004).

The GB women's hockey team is shown; this is the first women's hockey medal in 20 years. Memories are made of enduring time. Weight can be accorded through a breakthrough achievement. Every achievement shown is, in a sense, a breakthrough and part of the race-against-time, so it is hardly surprising that the language of hyperbole also deploys appeals to the extra-terrestrial, superhuman or timeless qualities of the successes. The performances of Brownlee brothers (Alistair and Jonathan), who achieved success in the triathlon by winning gold and bronze, are described as 'Team GB had delivered out of this world!'.

There is excess and there is honesty. Histories are made in sport through the creation and endurance of narratives of heroic figures of

DOI: 10.1057/9781137275363

honesty and those who 'play the game' within a discourse of honour. Just as some of the great moments of sport are tainted by dishonesty so are others framed by codes of playing by the rules. 'Being a sport' has itself entered the language as suggesting fair play. The comment 'Before the age of Usain Bolt it seemed the 100m was turning into a chemical alley' is clearly a reference to some of the pharmaceutical interventions which have enhanced performance in this and many other sporting events.

Everything is expressed in the language of the moment of crossing the line, the intensity of that moment – past and future are all condensed into one moment – and the seconds when victory is achieved, recognised and cheered. There is a sense of defeating time's winged chariot. Mo Farah appears; the 'history man'. On television the bell tolls in the last lap. The commentator says, 'does it toll for Mo', in the language of misplaced solemnity (or perhaps journalistic over excitement which generates malapropisms) which quite belies the exuberance of the moment. It all feels great now in the language of well being and immersing oneself completely in the present. 'The vibe is amazing! What an incredible experience'. Journalists and athletes seek to capture the immediacy of the moment in ways by which being there, whether virtually or actually, achieves more effectively, but by watching in whatever form we buy into what is both excess and constraint. The parameters of the event are set by the rules of spatiality and temporality as well as by the rules of the game, but the immediacy of the moment can generate affects which exceed any of these boundaries. The words of sports commentators are part of the event and the contract with followers, listeners and spectators, but it is the event and the assemblage of all the elements which makes it what it is and sporting events are made out of all these elements, in time, which is what makes them both so immediate and so exciting.

Eddie Butler, the Welsh rugby player and *Observer* journalist, comes on to endeavour to express these feelings; '2 weeks that changed the way we looked at sport and each other'.

Gary Lineker, who has temporarily transferred from his comfort zone of the boys' own world of the football league, to participate in the commentary along with women as authoritative as Clare Balding, on what is the much more gender inclusive zone of the Olympics, says some of the stories of 2012 will stay with us forever. To support these claims the coverage moves to vox pops backed by 'Maybe it's because I'm a Londoner'

DOI: 10.1057/9781137275363

to challenge resistance (especially as has been expressed in east London to the infrastructural disruptions of the games) in the context of legacy. The commentary endorses the local support for the games, linking temporality and spatiality; only a Londoner can answer what's great about the games – 'us Londoners'.

The build-up to the ceremony reinforces the immediacy of the moment by stressing on the party spirit and the great atmosphere, which focuses so powerfully on athletic achievement and celebratory social inclusion. This preamble is appropriate as a lead-in to the actual closing ceremony, which is a party of popular music, not all of it of the calibre of excellence which the athletic performances of the games themselves have achieved. There is little about actual legacy in either the preliminary review or the closing ceremony itself and little about the politics of the games, although there is discussion of the transformations of sexual politics which are manifest in these games. Women's participation and sporting success is so evident that the review cannot ignore it. 'These were the Olympics of the women' is a statement that is reiterated, and at that moment the image of a woman putting a hand with painted nails into her very brief knickers fills the screen, albeit briefly, but nonetheless suggesting some endurances. The disruptions are not as even as might have been thought or hoped. Preferred memories are being made but there remains some slippage between stated claims and the images which support or contradict them.

Making sense of memory and memories

The legacy of the games is expressed in the production of collective memory, which is also apparent in the regimes of truth which are constituted to locate the interest and engagement of the wider public, in the UK and the wider global audiences for the Olympics and Paralympics. BBC coverage is relatively local in the memories it makes and has a particular focus on London. Although the BBC coverage is officially marked by impartiality, it is not surprising that in this instance the focus is upon London and the collective memory is global in its reach but it is made as London's legacy and the games are both the 2012 Olympics and Paralympics and the London games.

Memories involve a mix of personal psychic investments and attachments, local loyalties and collective memories enacted in different levels and on different scales. Collective memory operates at different levels, which Zerubavel describes in the following way:

> collective memory is more than just an aggregate of individuals' personal memories, and such inevitably personal relief maps cannot possibly capture what an entire nation, for example collectively considers historically eventful or uneventful. (Zerubavel, 2004: 28)

This might devalue the dependence which media coverage places upon vox pops and individual narratives but nonetheless demonstrates the wider social application and significance of memories constructed at different levels in providing time maps.

Bergson defines perception and memory and access to both the present and the past in operational terms so that the present is that which is acting and the past is that which no longer acts (Bergson 1991: 68). Memory and perception are both geared towards action but are qualitatively different. Memories are not just faded perceptions and recollections but impel us towards consciousness, whereas perception propels us towards action: 'the past is only idea, the present is ideo-motor' (1991: 68). Perception is directed towards action and objects. Perception harnesses memory in order to make action possible. This applies to sports spectatorship and the process of making sense of the events and the spectacle itself. In the examples I have cited, it is apparent that one views each competition through the lens of previous performances and expectations in the active and fluid process of making and drawing upon memories and of envisioning the past through the present, and in this case the present through the past.

This approach differs from Merleau-Ponty's understanding of the elisions of mind and body (Merleau-Ponty, 1962) in that the focus shifts from the relationship between body and mind to one between mind and perception on the one hand and matter, that is the material which is perceived, on the other. Perception is central to the making of memories in sport as I have argued in the example of the 2012 Games in this chapter.

Bergson's work fits uneasily into most philosophical frameworks but has recently received considerable interest, especially by Deleuzian scholars and followers. Bergson's work, for example on matter and memory (1991), has application to the wider study of temporalities and has resonance for sporting times, much of which are concerned with memory. Bergson's vitalism counters the rationality of the Enlightenment and permits an understanding of change and becoming and of embodied development,

DOI: 10.1057/9781137275363

memory and especially for my purposes in exploring the importance of the making of memories through the present, of intuition and of irrationality, which does sometimes mark the intensities and expressions of the sporting moment. This approach permits the one I am taking in this book, which is seeing time as being in time and immersed in the moment, rather than, as Bergson says of space, which he distinguishes from time, the realm of rational objectivity (Bergson, 1991, 1921). Time is not amenable to rational evaluation according to Bergson, in a claim which sits uneasily with some of the concerns of the next chapter with measurement, but which nonetheless fits well into an understanding of memory as an intuitive construction of the past in the present. Whereas rationality is foxed and immobile, time is not. This approach has some resonance with aspects of phenomenological approaches which stress the lived experiences of temporality. As Bergson argues, 'we do not think 'real time', we live it' (Bergson, 1983: 48–9).

As much of the discussion in this book has demonstrated, sport is concerned with movement, action and perception and focuses upon the active participation in and experience of the moment, now. Memories are made through these movements and mobilities. These processes are active involving a movement forward; we can look back, but that assumes that there was forward movement prior to the looking back.

Time is thus an enduring past which unfolds and accumulates through memories. The past is created and in and through the present under the guise of a discernible set of certainties, such as the records that so frequently are invoked in sport but which are not necessarily constitutive of time and temporality, although these records and measurements use its properties to establish certainties.

Conclusion

Memories matter in sport in different ways, but one of the most distinctive features of sporting memories is the direct connection between perception and that to which perception is directed. The memories invoked in sport, especially as illustrated by the Olympic Opening Ceremony as in 2012, are powerfully enmeshed in social and cultural forces which further support the argument of this book that time is social. The relationships between past, present and future are mobile, whatever attempts are made in popular rhetoric to fix what is past and to embed the past in the annals of history.

DOI: 10.1057/9781137275363

Memory combines past, present and future through movement and through the social components of memory. Memory is social and cultural and these are forces which provide the dynamics of the relationship between different points in time. This can be an uneven process however and the euphoria of the moment in the present such as August and September 2012 may soon be lost only to be revived at the next ritual rehearsal of the greatest show on earth in Rio. Memories are constitutive of the excitement of sport because it is through memory that the event generates emotional affects. It is because we know the rules and what matters, what has happened already and speculate on what might happen next – now – that sporting events create such enormous attachments and enthusiasms.

Memory is also a process of engagement with what is real in relation to the present and the reality of experience which is strong associations with consciousness in which ideas are perceived before being represented. The operation of memory in the discussion in this chapter presents a different view of the relationship between mind and matter, especially between mind and body from Merleau-Ponty's phenomenological account but addresses some of the same issues which are particularly pertinent to an exploration of perception and memory in sport.

DOI: 10.1057/9781137275363

4
Measuring Time

Abstract: *Chapter 4 explores the technologies of measurement and the enfleshed experience of time and of losing time: embodied time and recorded non-human time. It explores questions about how the experience of being in time can be recorded or accessed and how different systems apply to particular sports. The chapter examines some of the ways in which the movement of time is recorded in media coverage and press and news reports as well as in the records of time measuring devices. Technological advances have led to ever more precise and sophisticated measurement of time which has influenced and shaped the techniques of athletes and their training regimes, strategies and body practices – body practices and recording mechanisms interrelate.*

Women's boxing in the Olympics in 2012 offers a most dynamic set of processes in which temporal forces are implicated in diverse ways: being in time and making history by changing times at the right time and where time is both a constraint and a liberating force. Boxing also presents an interesting example of the entangled forces of flesh and time and the measurement of time by the clock and through, for example, comparisons of the women's and the men's events in terms of times and timing. This exploration of boxing leads into the next chapter which engages with future time.

Woodward, Kath. *Sporting Times*. Basingstoke: Palgrave Macmillan, 2013. DOI: 10.1057/9781137275363.

 DOI: 10.1057/9781137275363

Much of the excitement which comes with sporting times is either seeking to push oneself, especially in athletics, to go faster or to witness others do so. Times have to be recorded; perception of speed is not enough either for athletes or for spectators. We seek the authentication of technologies which measure time, even if, on occasion, these findings are challenged and disputed. Speed is of the essence. There is nothing like breaking a record and going faster than everyone else for gaining accolades in athletics.

The experience of being in time may have the properties of fluidity and mobility as suggested in the previous chapter, but times as recorded for athletic achievements seem to be ever more precise and measured to achieve the highest degree of certainty. Memories flow to and fro and the past is re-envisioned in the present, looking forward to the future, but there is another aspect of time and timing in sport which seeks the security of fixity and a high degree of certainty. Indeed, in organised modern sport, timed records of performance are part of the definition of the set of activities that constitute sport (Guttmann, 2005). Bodies may be moving, often very fast and over a short distance, and the measurement of the time each activity takes is bounded by the certainties afforded by the apparatuses of measurement which have achieved greater precision than ever before in recent years. These claims however reveal nothing of the experiences of competing in time, or anything about the social and cultural implications and politics of measuring time. The fixity of measurement suggested by these statements highlights the centrality of time in sport and demands further exploration of the preoccupation in sport with these technologies of time and temporality.

This chapter explores the technologies of measurement and the enfleshed experience of time and of gaining and losing time: embodied time and recorded time. It explores questions about how the experience of being in time can be recorded or accessed and examines some of the ways in which the movement of time is recorded in media coverage and press and news reports as well as in the records of time measuring devices which have developed with ever greater degrees of precision (Frank, 2011). Different sports deploy different systems for measuring time and for establishing the criteria of success; in some instances succeeding requires coming first, second or third as in qualifying heats for most athletics, and in other cases, as in swimming, the measurement of time is what is used to determine which competitors enter the final rounds.

DOI: 10.1057/9781137275363

Timing is vital in the display of modern sport. The scheduling of the Olympics as the biggest of the mega spectacles is massively complex (IOC, 2012), but timing of the events themselves is crucial at all levels of sport. Modern life demands adherence to timetables and it is not only spectatorship which would be impossible without careful and detailed attention to timing of activities and events; participation is possible only with agreed timetables and schedules for reserving facilities and venues.

Time and times vary across sports and there are different internal systems which are used to regulate competition. There are different systems as well as different classifications of how time is allocated and monitored. There is extra time, time out, overtime (as used in basketball to resolve a tie) or even sudden death as an alternative means of limiting time (when the team to score first in the extra time allotted after full time has been expended is accorded victory). Full time, for example the 90 minutes allowed in football (soccer), is not finite. Ties demand time management, extra time at the end or some strategy to permit a result. Full time can be expanded to allow for injury time or stoppage time, when players take time out and additional time is added at the end. In rugby the clock is stopped for injuries. Halftime is the break in the middle.

In this chapter, women's boxing in the Olympics 2012 offers the most dynamic set of processes in which temporal forces are implicated in diverse ways; being in time and making history by changing times at the right time and where time is both a constraint and a liberating force. Boxing also presents an interesting example of the entangled forces of flesh and time and the measurement of time by the clock and through for example in comparisons of the women's and the men's events in terms of times and timing. Boxing in 2012 has implications for future time and the promise of change, projected into the future. Ironically, given the extent to which the sport has been embedded in traditional hegemonic masculinity and its apparent intransigence, this is the sport that offers transformations, where the potential for change is one dimension of future time. Future possibilities, progress and challenge to existing records and achievements are built into the measurement of time in sport. The precision of calculations in the present holds out the promise of future records. Measurements are relative in order to demonstrate unequivocally who has won and achieved, for example medal qualifying positions now, but also in the setting of records there is the possibility in the future for even faster speeds and better times, distances and techniques.

DOI: 10.1057/9781137275363

Sporting times; different sports

On the 27 of July between 11.23am BST and midday I watch the 400m freestyle in which for team GB Rebecca Adlington is the big medal hope. Becky is way ahead. I expect her to win after her success in Beijing and am excited to see her so far in the lead. Becky wins by 2 seconds. Easy, I think, caught up in the exhilaration of the moment. Now she is victorious, and I feel triumphant too (I am a fan of this woman who is a 'local girl' and lives not very far from where I do in the north of England and have followed her progress keenly), but swimming is different from athletics. There are four more heats and qualification is based on time and not order. Becky may have won easily but, after another four heats, Becky only manages to qualify in eighth position. Other competitors have achieved faster speeds. Again the measurement of time is central to sport but time and records become mixed. Time is against the clock.

Cycling time offers different recording mechanisms and systems. I am watching the women's cycling on television at 3.35pm BST on Sunday 29 July 2012. On this occasion, time is in the tactics. I am getting confused about who is winning this time. In the swimming heats it is clear who is in the lead, even if ultimately I find out the system of measurement means that the person who wins a heat goes through only if they are within the top-time categories. Cycling demands concentration and not only from the viewer; the top three cyclists need information too because time is knowledge and knowledge includes time. The leading three cyclists, Dutch Marianne Vos, team GB's Lizzie Armistead and Russian Olga Zabelinskaya, need to know the time split, that is the time between them and the peloton, the mass of other competitors bunched together behind the leaders. At least the television viewer can see where the peloton is, camera shots permitting, but the leaders clearly need to make swift critical judgements on the move. Time is more crucial than distance although it is distance which, albeit as an approximation, is what is available to the viewer – distance and space is visible whereas the calculation of time requires expert knowledge and even mechanisms and technologies through which an accurate set of information can be gleaned and then, as the commentators claim, be transmitted to the cyclists. This is more pressing for team GB as Lizzie Armistead is reputedly a 'fast finisher' and has to make a decision about when to attempt her burst. With or without the information, the outcome of the race is silver for Lizzie Armistead and gold for Marianne Vos and, in the post-match analysis, Armistead

DOI: 10.1057/9781137275363

good-naturedly muses over whether she should have attempted an earlier burst. The whole project is framed and governed by the specificities of time and its calculations, without which the high specification equipment, years of training and superbly athletic, fit bodies are found wanting. Time is much more than measurement and final outcomes.

Systems of measurement are complex, for example, in cycling. On the 2 August 2012, at 5.40pm BST I watch the men's team sprint. There is one team on the track and each performance is timed. Again this is not against other teams. At first I am puzzled; one man leads and then moves to the back. Time replaces team competition and tactics are internal but in the finals there are two teams and each can see the other, but on the different track. This is not simply one team versus another and seeing the competition ahead or behind you; they're on another track and going in different directions separate all the way. Maybe the cyclists can sense where the opposition is but as a spectator I am confused. 'Being in time is sport' is clearly constructed by the rules and regulations of the game which become the enfleshed materiality of its performance. Calculation in relation to competitors would be impossible in a sport like time trial cycling as the routine practice of the sport includes time and timing in its iterative body practices.

Sport engages with the internal ties of time and its constraints and the impact of time unrestricted by chronology. The measurement of time is subject to attempts at manipulation as well as classification, and the management of time through the technologies of measurement is not always transparent. Some interventions aim to reorganise the experience of time, even 'real time', now.

Manipulating time; re-situating experience

At 21.28pm BST on 30 August I watch the men's wheelchair basketball, GB v. Germany on Channel 4. The event is dynamic and fast-moving and incredibly exciting. The telecast stops for adverts 3 minutes from the end, which also seems incredible to me. How can the affect of so fast-moving an event be suspended, while products in which I have little interest are promoted? Time is to be adjusted so that we watch the last 3 minutes after the adverts in what is then recorded 'real time', but live to us viewers but not live to those at the stadium. Live time is recorded time! Commentators reassure us that we're going back to the fourth quarter and it'll be

DOI: 10.1057/9781137275363

exciting – is it? Especially after inappropriate advertisements promising us Paralympic excitement (which are not appropriate and are experienced as irritating) because we are already in a phenomenally exciting immersion in Paralympic sport, if only they would switch back to the games. The trailers persist while the real game is happening out of sight. Time is commercially managed. Play resumes and we are back in it but this process of time out is unsettling. Time out in the rules of the game is something else.

At 21:42 pm BST there is time out to talk tactics, even with only 3 seconds left. Tension increases. This time out is part of the event. The result is a draw with 66-all so it is all worth it. Now there is more time. This is 'over time' of 5 minutes; not the extra time with which I am more familiar as a follower of football. Play goes on at an unbelievable level of intensity until someone wins. The bursts of 5 minutes additional play will continue until there is a result. Germany forges ahead and wins after 5 minutes overtime: 77 to 72.

Time is recorded and classified in different, specialised ways. In the above example of basketball time out is when the clock stops and play ceases for a short period, which is 'overtime'. Time and timing are subject to different sets and regulatory frameworks as well as different specific nomenclature. Time is classified, coded and is also manipulated. It appears that nothing can stop time and the march of time but media systems and processes manage time and so, of course, most importantly do athletes, quite legitimately through tactics.

On 1 August 2012 at 7.30 pm BST I watch an interview with swimmer Michael Jamieson: 'It was more tactical. I forgot about the time' he says when interviewed after the race. There is some elision between tactics, which may be so much part of the athlete's training regime, and body practices that they are not thought and conscious but part of being in time (Merleau-Ponty, 1962). Cognition is always an embodied perception of the world in phenomenological accounts which provide a means of exploring the relationship between experience and knowledge through lived experience. The concept of lived experience, especially as translated in feminist accounts which focus upon the body as a situation as well as the body as situated (de Beauvoir, 1989 [1949], Moi, 1999, Young, 2005) permits an understanding of the situations in which knowledge is produced and how knowledge and the producers of knowledge are situated.

These ideas about embodiment which includes the merging of mind and body and a development of Pierre Bourdieu's idea of *illusio* have been conceptualised by Löic Wacquant in his work on boxing in order to

DOI: 10.1057/9781137275363

explain some of the ways in which boxers persist in the ring, even when they appear to be injured and would be expected to be in considerable pain, in a manner which denies the possibility of reflection and is 'in the blood' (Wacquant, 1995, 2004). This is less a tactic than a way of being. It might however be part of the iterative training practices which make body practices so routine that the athlete does not have to think about them. Strategies of managing time too could become routine at the most elite levels.

In some sports there is more scope for conscious reflective strategies. For example, in swimming there are two approaches to managing time. First, time means going for the best time and second there is tactical time which involves competing against others in the pool. Tactical time, of course, applies to all sports although the sets of practices which are deemed appropriate vary as do the levels and extent of conscious decision-making.

On 2 August 2012 at 11.50am BST, I am watching the men's double sculls. Expectations are high as Cohen and Sullivan broke the Olympic best time in the heats. It is an exciting race, largely because the New Zealanders surge dramatically at the end, storming past Italy to take the gold medal. 'This one race, this moment in time' says Garry Herbert, commentating on the men's double sculls (heavyweight) final and on Joseph Sullivan and Nathan Cohen coming in first and winning gold for New Zealand. 'In harmony with your boat', says Redgrave. 'They had no idea where they were'. 'The Italians knew where they were (2nd)' says John Inverdale, 'You Can't go flat out all the time – you have to conserve and expend energy strategically'. The Italians had a burst in the middle of the race but couldn't maintain it for the final burst and New Zealand got through. 'Rowing is all about pacing', says Steve Redgrave.

There are tactics and practices which form part of the rules of the game and such practices change according to changing times and different informal as well as formal codes as the former cricketer and sports writer Ed Smith argues using his own experience of the changing codes of honour in cricket (Smith, 2008). Time is implicated in these rules and regulations formally, for example as set in the Olympic Charter (IOC, 2012), or in the very practices of sport.

Rules and regulations

Modern sport is highly organised by a whole set of rules and regulations, not least the Olympics (IOC, 2012). Not only are these rules carefully set

DOI: 10.1057/9781137275363

out and detailed, they also change on a regular basis to accommodate social and cultural changes and political considerations, such as widening participation in sport and promoting greater equality and democracy. For example, the inclusion of women's boxing in the games for the first time as a competitive sport in 2012 resulted from a decision made in 2009 to amend the Olympic Charter. All sports are rule governed with written and/or unwritten rules. Even an informal kick-about in the park requires participants to follow the rules; this is part of the definition of sport (Guttmann, 2002, 2005).

Rules are often embedded in the body practices which make up the sport so that they become taken for granted. Some sets of rules, including those which focus upon body practices, are so complex that they demand sophisticated techniques of measurement to ensure compliance and a detailed grasp of the details. Umpires and referees are increasingly supported by a wealth of technical assistance and visualisation and sound technique to assess whether or not a body is over the line or has touched the ball or the wicket or is offside. Bodies are not the only subject of sporting regulations, which are also concerned with space, it also includes distances and the size of the pool, court or pitch, and time.

In the Paralympics, bodies are the target of the classificatory system, which categorises athletes by the impairments which are deemed to lead to competitive disadvantage. Formally this means that the categories are designed to reduce the limitations to performance and ensure that success is determined by skill, fitness, power, endurance mental focus and tactical ability (Paralympics, 2012). Time is built into the concept of endurance although times are not usually part of the formal framework of classification. These classifications however are concerned with the possibilities of change and are themselves constantly subject to modifications which have significant temporal implications. The more sporting regulatory bodies seek to establish boundaries and some certainty and fixity, the more social and cultural changes call into question the constraints of regulation.

Athletes are grouped into classes defined by the degree of function presented by the disability. There are four broad categories: athletes with an amputation, who are called 'amputees'; athletes with cerebral palsy; athletes with visual impairment or blindness; and athletes with spinal injuries or other physical disabilities.

London 2012 compares the classifications of disability to those between women and men, which may or may not be helpful to spectators. Gender

DOI: 10.1057/9781137275363

classifications, especially as evident in the controversial gender verification testing practised by the IOC along with other regulatory bodies like the IAAF, are highly controversial (Woodward, 2012a).

> If you are new to watching Paralympic sport and classification, it can help to think of why men always compete against men and women almost always compete against women. Or think of it in similar terms to how boxing at the Olympic Games works where athletes compete in different weight categories such as Flyweight or Heavyweight. (London Paralympics Classification, 2012)

These classificatory systems are complex, especially in the numerical labels which are used to describe each of the 20 different sports which are included in the Paralympics. The classificatory system is regulatory but explicitly designed to generate fair play. Nonetheless the more complex it becomes, the more controversy is created. The concept of fair play and the equal playing field for athletes with different disabilities may be aimed at providing an equitable and fair regulatory framework for elite disabled athletes and comprehensible rules for spectators, but the more sport involves interventions aimed to enhance performance and to increase speed and achievement, the more difficult it becomes to draw boundaries around body practices which are not implicated in economic social and cultural practices.

Time and its precise measurement through the development of scientific mechanisms and apparatuses of quantification (Shallis, 1983, Sheely and Bourgeois, 1994) are built into the regulatory technologies and systems which govern sport. Sports are temporally circumscribed as illustrated by designated time periods, such as halves, quarters and innings, or number of bouts, games, sets or rounds or, as has been shown in this chapter, with diverse body practices which seek to negotiate and manage these temporal restrictions. There are a number of adaptations to prevent indefinitely long sporting encounters. Some sports have short versions, like one-day cricket with 50 overs, Twenty20, Day-Night matches in cricket and, of course, in the more explicit synthesis of bodies and rules through sexual politics, and the shorter times allotted for women's competitions. Also, as the examples of the 2012 Games show, there are a variety of strategies which have been formally instituted to deal with draws, such as extra time or over time.

The processes of surveillance and regulation through which sport is monitored and practised include a diverse range of organisations and

DOI: 10.1057/9781137275363

infrastructures, some of which relate to the commercial sponsorship of sport and the bodies that make up sport as a globalised and highly profitable industry. The Olympics are no longer, if they ever were, a sporting event set apart from the pursuit of profit and the cult of amateurism (Guttmann, 2002, Preuss, 2008). Sponsorship and the impact of increased professionalism in Olympic sport involves much more than the commercial logos which are displayed in the Olympic Park and the use of the Olympic Rings and the London Organising Committee for the Olympic and Paralympic Games, LOCOG, (LOCOG 2012) branding. Different games have different sponsors, but one of the continuities in the presentation and representation of the games are media infrastructures through which the event is promoted, and in the case of the Paralympics 2012, it includes a commercial broadcaster as well as the BBC with its restrictions on advertising. Time is also money in the period of time permitted for the promotion of products and services.

Other ways of managing and manipulating time are part of the framework of sponsorship and promotion of sport. Viewers of commercial channels, which are virtually all broadcasters except the BBC, are familiar with interruptions to their concentration and the disruption of emotional affects with frequent advertisement breaks. Sports fans are more used to some respect for the timing of their sport and for carefully positioned advertisement promotions; at half time or at the end before extra time if this is required in football. In cricket, the system works well for the dedicated fan; they are likely to be at the end of an over with a single advertisement or if someone is out there could be time for a few. Take for example, the Paralympics on Channel 4, which is a commercial channel and is subject to the interventions of advertisers that interrupt the flow of competitions (as illustrated above in the case of the men's wheelchair basketball). The interruption in this case is a bit more troubling for the spectator as excitement mounts with high, very even scores. It is not just the swift scoring or the tension of two teams competing so closely. The embodied practices of wheelchair basketball are incredibly swift and exciting. Competitors topple out of their chairs only to be reinstated quite undeterred at high speed and the spectator is drawn into the now of the event. Then we move to the advertisements again. This is cruel. There are only a few minutes of play left. When play resumes, the clock has technically stopped. We pick up where we left off.

There are other interventions and types of manipulation from the sponsors of sport and those who supply sports equipment. Other

DOI: 10.1057/9781137275363

practices involve equipment and clothing which can enhance speed and efficiency in the performance of sport. Such interventions have been recognised and, for example, in swimming, the specialised clothing has now been regulated and new, less advantageous swimming costumes imposed. Specialised clothing and equipment can enhance performance and improve speeds albeit by fractional amounts of time, even a tiny fraction of a second, but it does make a difference.

At Beijing in 2008 there was considerable excitement about swimming successes of athletes wearing new, very hi-tech swimsuits. These were the first swimsuits to combine stitch-free, ultrasonically welded seams, water-resistant fabric and strategically placed polyurethane panels which had been designed by NASA. The LZR swimsuit made considerable impact at the Beijing Olympics. In Beijing, LZR swimmers, most notably American Michael Phelps, won 94% of the gold medals on offer and broke 23 of the 25 records set in the Cube. Opponents of the hi-tech suits argued the buoyancy they created amounts to 'technological doping' (BBC Fina Swimwear, 2012). The world swimming governing body, Federation Internationale de Natation (FINA), decided to limit the use of non-permeable materials in the suits (BBC Fina Swimwear, 2012) and intervened in order to contain the impact of these controversial hi-tech swimsuits. Swimmers are also no longer able to wear anything under their costumes after Swedish swimmer Therese Alshammar's world record was erased because she was wearing two swimsuits. An extra suit can help swimmers compress their body and trap air, which provides greater buoyancy in the pool. For example, of the 105 world records broken in 2008, 79 of them were by swimmers wearing one suit, the Speedo LZR Racer (BBC Fina Swimwear, 2012).

Alshammer, who improved her own record in the 50m butterfly in Sydney, was the first swimmer to have lost a world record under the new rules. Although she is reported to have claimed that she believed she was allowed to wear a 'modesty suit' under her skin-tight racing gear (BBC Fina Swimwear, 2012). The executive director of FINA, Cornel Marculescu, made it clear that nothing should be worn underneath the swimming costume. Arguments centred on the protection of their modesty which women athletes required and, for example, some officials had stated that women were allowed to wear bikini bottoms for modesty purposes at the event but not a full suit and any world records set would not stand if they used them. FINA as the international regulatory body of the sport has made it clear that swimsuits should not cover the neck

DOI: 10.1057/9781137275363

and must not extend past the shoulders and ankles. These changes have also been imposed to limit the thickness and buoyancy of the suits.

FINA is reported as stating that the swimming costumes which had been brought in constituted unacceptable performance enhancement: 'FINA wishes to recall the main and core principle is that swimming is a sport essentially based on the physical performance of the athlete' (BBC Fina Swimwear, 2012). This example is cited to demonstrate some of the complexities of the entanglements of enfleshed athletes, regulatory bodies, cultural practices, economics and social systems in the practice and performance of sport, all of which aim to increase speed.

Body practices and timing techniques

Engagement with time and timing can be seen as one of the body techniques which need to be developed in order to compete as an athlete. The example of the time trials in cycling brings together two important features of temporality. First, time is social in that it is influenced and the experience of time is in part shaped by the rules. The body practices or techniques, as Marcel Mauss called them (Mauss, 1973), are performed through a body which is a technical instrument, which is organised in different ways in different cultures (Mauss, 1979), of which sport is one aspect. Second, being in time brings mind and body together in body practices, in what R.W. Connell called bodily reflexive practices (Connell, 1995) which become routine. Sport is all about developing these techniques and body practices and I suggest time and timing is implicated in body practices which can usefully be understood as bodily reflexive practices. Connell argues against excessively social constructionist approaches to bodies as socially inscribed in an approach which underplays the materiality of flesh. Connell's approach uses the idea of bodily reflexive practices to explain the materiality of bodies which incorporate mind, body and social situation. Bodily reflexive practices are

> not internal to the individual. They involve social relations and symbolism; they may well involve large-scale social institutions. Particular versions of masculinity are constituted in their circuits of meaningful bodies and embodied meanings. Through body reflexive practices, more than individual lives are formed; a social world is formed. (Connell, 1995: 64)

In Connell's conceptualisation, bodies are both agents and objects of practice which are socially situated. The rules of the game and the practices of sport are made through its body practices. The materiality of the

DOI: 10.1057/9781137275363

flesh both enables and limits what the infrastructures of the social world can create. Thus the bodily reflexive practices of sport also are constitutive of gender differences, rather than those differences being entirely generated by the body the athlete inhabits. Rather than isolating social and enfleshed materialities they are seen as co-constitutive and part of the exchange between the promises of agency and intentionality and the constraints of culture and flesh. Within sport and the bodily reflexive practices of sport, temporality is caught up in this mix; keeping time and, especially, improving upon times, where possible to be recorded in sporting times, are part of the intentionality of athletes as well as being the targets set by the governance and culture of sport.

Connell's focus upon the gendered dimensions of bodily reflexive practices has application in sport, none more so than in the sport which has been most strongly associated with hegemonic masculinity (Woodward, 2006) and even the traditional embodied extremes of machismo. In 2012, things are changing, however.

Boxing time

9 August 2012 at 16.15 pm BST is the final of the women's flyweight boxing at the ExCel. Team GB's Nicola Adams is fighting Ren Cancan, the Chinese champion. In the semi-finals Adams has already defeated India's MC Mary Kom, who was strongly backed to win gold before the games. This is not just any fight; this fight is already a very important event. It is the first time it has been possible for women to compete in the Olympics. There was a display in St Louis in 1904, but although women were boxing through the twentieth century they were not at the Olympics. Women's inclusion was controversial so there is a lot hanging on this fight as one of the three weights included in women's boxing in 2012.

The first round is fast moving and clean. Adams delivers several clear punches. Two minutes pass very quickly. 4–2 to Adams. Round two and Adams lands a great left on Cancan followed by a right to the temple. Still only 5–2 but Adams is on top and looks good. Round three and Adams is dancing; she is small and agile but Ali springs to mind. She has said he is her hero; well she is emulating some of his moves and Cancan is having difficulty touching her. Going into round four it is 14–5 and Adams is still dancing out of trouble. Then the coup de grace – a left hook. This time it is Sugar Ray Robinson, I think of. He always finished off with a left.

DOI: 10.1057/9781137275363

Time also matters in boxing in particular ways and the measurement of time is central to the organisation of the sport and the structure of its enactment as well as the body practices of participants in the sport. In the Olympics, men's rounds each last 3 minutes and there is a total of three rounds per fight; in the women's game there are four rounds each lasting 2 minutes. Rounds are marked off by a short rest in between. The rationale for the different time schedules for women and men suggests that women need more frequent recovery periods and that women are able to sustain the effort of boxing for a shorter period of time than men.

The regulatory and disciplinary practices within the ring, which has a clear engagement with time and temporality, are stated by the IOC and according to the rules of amateur boxing as laid down by the International Boxing Association, AIBA (AIBA, 2012). A boxer is considered down if he or she, as a result of being hit the boxer, touches the floor with any part of the body except the feet. The referee counts from 1 to 10, 10 seconds. The count is now electronic and mark by audible beeps for each number although some referees still count themselves. The referee signals the count and if the boxer is still down after 10 seconds, it is a knock out. If the boxer gets up immediately after touching the ground there still has to be a count of 8 and after 8 seconds have passed the referee signals that the fight can continue by saying 'box'. Time and clock time in particular is used to provide fair play and to protect boxers from the risks and dangers of their sport; not that boxing is the most dangerous of sports (Woodward, 2012,Dangerous Sports, 2012). Speed may be a body practice in this sport which is used to display maximum threat to the opponent as well as to remove oneself from the danger zone of harm, but time and the control and regulation of time is also integral to the safety practices which make boxing a legitimate, and, as in the case of women's boxing in the Olympics, an agile, fast moving and very entertaining and sport.

However, the inclusion of women's boxing as an official sport in the games has not been without opposition (Woodward, forthcoming). There was considerable debate about the extent to which women might be more likely to suffer damage than their male counterparts and a rehearsal of the arguments used as far back as Aristotle in the ancient world and reiterated through the nineteenth century, especially by the founding fathers of the Olympics (Woodward, 2012a) who suggest that women's reproductive capacities could be impaired by over exertion (in Woodward, 2012a and forthcoming). Some of the opposition was

DOI: 10.1057/9781137275363

expressed in terms of anxiety about spectatorship which does have some substance, for example in relation to debates about pornification and the sexualisation of women's bodies and the risk of voyeurism (Woodward and Woodward, 2009).

The other discussion is about women's boxing being both out of place and out of time. Prior to the inclusion of women's boxing in the Olympics in 2012, there was some discussion about whether or not women should wear different clothing; skirts were suggested, in order to make it clear that they were women and not men, for the information of those watching fights on television at home. Clothing is one of the body practices through which flesh and social and cultural forces elide and is certainly a clear marker and maker of sexual difference. The regulating bodies of sport, in this case the IOC for the Olympics, deemed it necessary to create clear and distinctive markers of visible sexual difference because the body practices of the sport, with its techniques of punching, dodging, weaving and dancing, might not in themselves be sufficiently invested with the distinctions of sex gender. There is also the implication that viewers who casually switch on their television set and see boxing need to know that they are not watching men but women, most likely because of all sports boxing has been the one most deeply embedded in hegemonic masculinity (Woodward, 2006).

One of the most marked achievements and successes of the inclusion of boxing in the Games in 2012 however is that it was virtually impossible to distinguish between women and men in the ring, except by size at the extremes of the weight spectrum. There is no heavyweight category for women; in the Olympics there are just three weight classes, with the heaviest being middleweight. Perhaps a heavyweight category would be too unladylike or too threatening to men's boxing? Women's fights at all stages of the competition and in each weight class were enacted within the same frame work as the men's fights, albeit with fewer weight classes. The commentary, for example, on television and in the press on the Olympics sports pages was within the language of boxing as is particularly demonstrated by the coverage of Nicola Adams's gold medal and her fights at all stages through the competition.

One of the dimensions of time and temporality which is so well-illustrated in the sport of boxing especially in relation to sporting times is its legends, myths, heroes and histories. Boxing more than any other sport is classified not only through its regulatory bodies' classifications of body practices, times and spaces, as all are, but also by historical periods.

DOI: 10.1057/9781137275363

Boxing has golden ages and times of 'when we were kings' (Boddy, 2008, Mitchell, 2009, Powley, 2011, Sammons, 1988, Woodward, 2006, forthcoming). Women have boxed all through the nineteenth century when they had very little visible presence (Hargreaves, 1994, Woodward, forthcoming) and at the end of the twentieth century there were women who might have been deemed to have achieved the status of boxing heroes. Such women, however, were often exalted through their familial connections and categorised more through their famous fathers than their own, certainly not insubstantial achievements in the ring and in competitive boxing. The most notable were Leila Ali and Nancy Frazier. I probably do not need to say that they were the daughters of the Greatest, Muhammad Ali and 'Smokin' Joe Frazier respectively, such is the fame and heroic status of these giants of boxing legends, especially Ali. What has contributed to the exclusion of women from boxing and underpins some of the debate about whether or not women's boxing should be permitted by the IOC at the Olympics is not only the anxieties about the quantifiable damage and measured body practices in the ring and women's capacities or not to endure the corporeal stresses of the sport, but the desire to protect the heroic narratives in which so much of hegemonic masculinity is invested (Woodward, 2006). It is not so much body weight and embodied endurance and resilience which keep women out as the impossibility of putting women into the sporting times of boxing, its histories and legends. Boxing legends so often invoke the narrative of escape from the ghetto and the route which boxing has traditionally afforded migrant and oppressed men out of poverty and into self respect. This is not a romanticised view because this route has often been marred by corruption and exploitation within the sport and boxing has as many bad guys as good guys and its fair share of those who are both, the best known among whom must be Mike Tyson. However, the stories that have been told and the credibility afforded to boxing have validated and exalted the sport at pivotal points, hence the narratives of greatest and invocation of 'kings' as well as serving to exclude women, ostensibly on the ground of enfleshed practices and cultural references to the fragility of femininity, but more substantially, because the assemblage of diverse elements which has made boxing culture, including its heroic legends, impossible to put women into the mix.

Women's boxing in 2012 made it possible to think the unthinkable and to start telling stories about women like MC Mary Kom, Ren Cancan, Katie Taylor and most especially of all, Nicola Adams. It becomes

DOI: 10.1057/9781137275363

possible to locate women's engagement in the set of body practices and compliance with a set of sporting regulations within an established time frame and to describe their performance in the same discourse as men's engagement in the same activities.

What this does demonstrate is the ways in which different aspects of time and temporality intersect and combine. This includes the clock time in which so many boxing expressions of temporality are expressed, such as saved by the bell and out for the count, but which are also constitutive of its narratives and ways of constructing regimes of truth about who can and who cannot take part and achieve success. These are all part of the social as well as the enfleshed processes of being in time in sport.

Conclusion

This chapter began with an exploration of how time is central to sport through manifestations of the measurement of clock time and of the ways in which sporting time is measured and calculated. The measurement of time in sport, as in contemporary life, is not only marked by the speed of its passing and the pressure to achieve within a restricted time frame, but also by the development of ways of measuring and of negotiating these constraints.

The experience of being in time in sport suggests some insights into the experience of temporality which so marks contemporary life which is governed by the clock rather than seasonal and natural movement and change (Adam, 2004). Time and 'beating the clock' is represented as a target in sport but it is also a means of going beyond the present and of transcending 'now'. Transforming techniques of measurement are allied to new ways of negotiating time and the possibilities and promises of thinking differently, even in so strongly regulated and monitored a field as that of sport. This chapter has demonstrated some of the sometimes surprising and counter intuitive ways in which time is regulated and even manipulated, especially in relation to the subversion of expectations in the experience of spectatorship and in developing strategies in the field which facilitate maximum outputs.

Time includes clock time with stop watches and precision equipment becoming more and more deployed to establish records and resolve disputes in sport. Time is movement from A to B, but it is always social, albeit in different ways. The regulatory systems of sport offer specific ways

DOI: 10.1057/9781137275363

of measuring time in relation to the movement of bodies through space and suggest ways of negotiating time in the present. The measurement of time is closely linked to the now of performance and of spectatorship but it is always in conversation with what has been and with what might yet to come in the present, in the competition that is happening at this moment. Time involves more or less manipulation and illusion which demonstrates the importance of perception of time and the experience of being in time. It is difficult to attempt to measure the perception of time and even timelessness and a state of being. The future is not only the matter of going faster but it is also concerned with managing time through recognition of social time.

The next chapter looks at the future and at what future time might look like in sport, in a field in which the records which are measured, aspired to and surpassed might put restrictions on the future as well as holding out the promise of success yet to come.

DOI: 10.1057/9781137275363

5
Future Time

Abstract: *Chapter 5 looks at how the future and its promises are lived in the present in sport and at how the future relates to the present and the past in relation to sporting performances. Futures are explored in relation to predictions and possibilities, for example as they involve generation and aging and damaged embodied athletes. There are enfleshed restrictions of performance that constrains the promise of futures as well as technologies which create new opportunities, as in the Paralympics. Power and agency underpin these debates. The 2012 Olympics were at various points described as 'women's time' and, in the context of the Spice girls' reprise of their first hit Wannabe which purportedly introduced the notion of girl power, at the Closing Ceremony, the 'girls' games'. The enormous success of women's sporting activities at the Games and the operation of different power mechanisms based on divisions of sex gender in sport make the concept of 'women's time' an interesting political device for exploring temporality, especially in the field of sport, but with much wider application.*

Woodward, Kath. *Sporting Times*. Basingstoke: Palgrave Macmillan, 2013. DOI: 10.1057/9781137275363.

 DOI: 10.1057/9781137275363

This chapter looks at how the future and its promises are lived in the present in sport and at how the future relates to the present and the past using the example of sporting performances. Futures are explored in relation to predictions and in particular to possibilities, as they involve sex gender, generation and aging and impairment. There are enfleshed restrictions of performance that constrains the promise of futures as well as technologies which create new opportunities, as in the Paralympics. Power and agency underpin these debates. Future time embraces ever faster and more precisely measured records as well as the development of technologies and interventions which legitimately enhance performance, but this is not the whole story. The matter of legitimacy is itself open to future transformation and interpretation, when what is disputed and even prohibited in the present maybe become acceptable and desirable in the future. Change and futurity are concerned with more than simply reversing the present, predicting the future or dreaming utopian dreams, as feminist critics have demonstrated, albeit largely in empirical contexts other than sport. Future time is conceptualised in the present and thus relates to what is thinkable and to what can be imagined beyond the present. Part of this discussion relates to technological advances, but, as has been argued in this book, time is always social so that the promise of future time can be largely seen in the context of what is likely to endure and what might change in social relations and divisions as well as in body practices and the enfleshed actualities of sporting practices. Different axes of power intersect in the making and imagining of future time and future times.

Sex gender, lived and embodied, is a crucial aspect of what is social about time in sport (Woodward, 2012a), especially in its classificatory systems and in demonstrating some of the specificities of flesh and time. The life of any athlete as an elite performer is limited but time and its passage puts more limitations upon women than on men. Virtually every sport is classified by sex. Women run shorter distances and are assumed to take longer than men in any competition they enter. Embodied difference carries considerable weight, and weight is sometimes how it is measured, in sport; women even receive less media time than men in sport. The Olympics however are more democratic and egalitarian in many ways than several other sports, both in the equal participation of women and men and in the media coverage. Women's competitions receive almost equal coverage to men's and there are far more competitions which women can enter. The 2012 Olympics were at various points

DOI: 10.1057/9781137275363

described as 'women's time' and, in the context of the Spice girls' reprise of their first hit Wannabe which purportedly introduced the notion of girl power, at the Closing Ceremony, the 'girls' games'. More substantially, the Games were seen as the 'women's games' because of the massive interest which was generated by some of the women's competitions including the women's boxing, with Nicola Adams attracting a huge amount of attention and interest for her competence and skill and her engaging presence. On the peripheries there might have been some engagement with the freak value of women participating in what has come to be known as a man's sport in which the display of machismo is ever present countered by an expression of women's corporeal fragility, as I know having conducted several interviews with local radio stations. However in the mainstream, including serious press and the BBC for whom I also did interviews, and for the majority of viewers, women's boxing was great entertainment and Nicola Adams was the most likeable hero of team GB. These are ordinary times as well as the days of miracles and wonder (to borrow from Paul Simon's song, The Boy in the Bubble).

The enormous success of women's sporting activities at the games and the operation of different power mechanisms based on divisions of sex gender in sport make the concept of 'women's time' an interesting political device for exploring temporality, especially in the field of sport, but with much wider application. It also opens up possibilities and routes into conceptualising the future, not as a target to be met with quantifiably more women participating in sport and especially in particular sports, but also new ways of thinking about sport and its body practices and new ways of being in time.

Time and the human condition

Time is however an irresistible and irreversible movement forward which, although it is complex and divisive, has only one trajectory or direction. Whatever hopes, dreams, transgressions and promises there might be, there are also constraints and a bottom line of extinction for human beings and organic matter which makes time and its limitations a serious matter. This is evident in the attention which has been paid to time's forward march as an inevitable aspect of the human condition in the myths and legends and literary and artistic attempts to engage with time and temporality. The active force of time is also intricately tied up

DOI: 10.1057/9781137275363

with pleasure and the pressure to 'seize the day' and make the most of what time we have. In sport there is an even more limited period for the enjoyment of some pleasures, notably those of elite athleticism, than other areas of experience. As Andrew Marvel stated so powerfully in his poem *To His Coy Mistress* (Marvel, 1997); 'But at my back I always hear/ Time's winged chariot hurrying near.'

This poem focuses on some of the temporal conflicts between haste and languor, fear and pleasure and is constructed around the tension between the slow, languid lingering-in-the-moment, which is usually in contrast to the energies of the immediate moment in sport, the enjoyment of the present without thought for future decay and disintegration, and fear of the speed with which failure and collapse and ultimately death are always approaching. Pleasures, especially those associated with sexuality and bodily desires, offer some resistance to 'time's winged chariot' and suggest that the body and pleasure can be forces of resistance as well as being the source of human frailty. Although informed by the promise of seduction, the poem expresses the ghastly human dilemma of time. There is never enough time and human life is finite; 'the grave's a fine and private place/But none I think do there embrace' (Marvel, 1997).

The way in which temporality and pleasure are interwoven and inextricably enmeshed is part of the social and enfleshed experience of sport too. Pleasure is deeply embedded in sport and sporting times; sport retains some of its playful dimensions although it has become a significant force in global capital. The Olympics generated some elements of party and partying, which was a key feature of the closing ceremony. Pleasure in sport related to the experiences of participants and of spectators is given expression in relation to time in the transcendence of consciousness of time and in timelessness and the possibility of peak performance taking those who engage in the corporeal activity and those whose senses are implicated in spectatorship into the zone. The promise of pleasure is also part of the imaginary of futures. This chapter explores some of the possibilities of imagining futures which incorporates pleasure and the capacities of sport for imagining what is yet to come and provide some alternative approaches to futurity, beyond setting targets or making predictions. Pleasure is also, as Foucault demonstrates, aligned to bodies and to power (Foucault, 1981).

Foucault argues that there is a close association between the body as the locus of operations, on the one hand, of power which can be constraining and repressive, and, on the other, of resistance. Pleasure is what

DOI: 10.1057/9781137275363

induces bodies to participate in power (Foucault, 1981). In Foucault's work the pleasurable forces of the body are expressed through sexuality, which is one of his prime concerns, but sexuality is an amalgam of body pleasures, desires, fantasies and ways of being which have implications for the enjoyment and representation of bodily practices and pleasures which may be outside the more commonly held parameters of sexuality. Thus the normative disciplinary operations of power which characterise contemporary social life including sport and the regulation of sport's practices and networks entwine themselves in sporting pleasures. Sport offers a convincing illustration of the mechanisms through which pleasure functions in the service of power, whilst pleasure too consolidates resistance. Whilst Foucault places some emphasis upon the resistance possibilities which pleasure affords, pleasure does not have to be political although pleasure may generate political affects (Grosz, 2005).

Pleasure is one of the components in the assemblage of power, affects and bodies. Another aspect of focus upon bodies and embodiment is the relationship between inside and outside; the orientation of the body to the outside world and its own inner states, including psychological states and the question of consciousness. Much of the discussion in this book has focused upon the relationship between seeing and being seen, between performing and spectatorship and between perception and the objects of perception in time. Bodies are central to the idea of subjects who are not simply free agents with consciousness who exist independent of perception and action, but extend into the world through real and virtual actions enacted temporally as well as spatially. Thus resistance and change are possible only through the creative combination of perception and corporeality, which is material, that is mind and bodies are involved and perception and consciousness are not separate and discrete nor are they individual and unique. As Bergson suggests

> If our body is the matter to which our consciousness applies itself it is coextensive with our consciousness, it comprises all we perceive. (Bergson, 1977: 258)

Thus the temporal sensation of 'real time' combines all these forces.

Future time: transcendence

On 1 August at 11.50am I am watching the 2,000m women's pairs. In particular I am watching Helen Glover and Heather Stanning of team GB for partisan reasons and I am captivated by the fact that they are way

DOI: 10.1057/9781137275363

ahead with the promise of gold as the excited commentator keeps assuring us as I watch on television. I am also overwhelmed by the way the two women, the boat and the oars cut through the water; this is not just synchronised rowing – they all seem to be unified in the process of slicing through water in a forward movement. They are so far ahead that gold indeed looks very likely; Glover and Stanning are lengths ahead rowing in perfect harmony. What really draws me in the most spectacular and sensational manner is the way in which bodies, oars, boat, water and motion are in harmony and completely synchronised. The excitement in the crowd mounts, the commentators shriek euphorically but none of it is in any way part of the bodies in the boat; the bodies in the boat belie the ecstatic cries and the mounting hysteria about team GB's first gold medal. They seem to be floating quite unaware of their surrounds in which cries of gold medals, records and quantifiable achievements pour forth. There is an elision of all elements in something more than an assemblage; it all seems to be merging in slow motion even though Glover and Stanning are going very fast, much faster than anyone else in the race. It is the embodied harmony of the two women, so in time, who are seemingly oblivious to the shouting and are caught up in the movement and the process and appear to be going beyond it, beyond discourse and beyond representation even though I am watching, transfixed. At the end, the two women's bodies are entangled in exhaustion and only slowly part and become two separate embodied athletes. I know they are exhausted and I have seen this slow emergence from the state of physical exertion before, especially in rowing, when exhausted men lie back or lean forward with their head between their knees, short of breath, in recovery, but not like these two entangled bodies lost in time. This time it is going on for ever and I am caught up in the timelessness of the moment too. They keep on touching, being together and being a team, being one and being two in time. The language of hyperbole and terms such as amazing and incredible, however discredited they are by over use during the games, spring to mind and I am sorry that there are insufficient lexical items in my repertoire to capture this moment in 'real time' which may well, of course, be beyond discourse. What does spring to mind is Luce Irigaray's feminist version of *jouissance* (Irigaray, 1984, 1991) as that pleasurable state of pleasure, for example between mother and child which is outside and beyond patriarchal language.

This looks to me as an event which situates me outside and at a distance and yet drawn into the experience, unexpectedly and inadvertently, like

DOI: 10.1057/9781137275363

a moment, in fact quite a long, immeasurable moment of being in the zone into which the spectator is also drawn. It suggests the social and cultural applications of being in the zone which is palpable as well as observable. There are pairs of performers who are caught up in a process of being one as a manifestation of being in the zone; ballet dancers Margot Fonteyn and Rudolf Nureyev or even ice skaters Torville and Dean whose synchronised performances also exceeded technique and athletic, enfleshed expertise. What makes this occasion memorable is the qualities of the experience after the competitive event rather than the event itself. Although the metaphor is spatial, the transcendent properties of being in the zone seem more temporal than spatial in this instance and less related to the athletic performance itself than to its enfleshed aftermath and recovery. This offers a new way of looking at future time which extends far outside the future as a set of targets and objectives or as a time to predict. Neither is it a utopian future because utopia seems to be now. Futurity is both within the present and beyond it.

Bodies in time: enfleshed possibilities

In the above rowing example the nature and quality of the enfleshed encounter is what makes the event distinctive; this goes beyond the body practices and strategies of sport which are what makes bodies central to sport and sport so exciting. Bodies are central to an understanding of time which makes sport a productive site for the exploration of time and temporality and the potential and restrictions which time places on us. Thus it seems to be the most useful and creative way of understanding temporality as going beyond discourse which suggests alternative routes into the future, but which nonetheless demands that we are attentive to the ways in which bodies are implicated in the process.

This is about experience, but it is not limited to experiential accounts especially since, as I have suggested, the hyperbole of commentary and even of language may fail to capture the whole experience. The methodological engagement with 'real time' which I have adopted in this book does require accounts of experience, however limited. The methodological approaches to 'real time' which are adopted in this book are experiential and concerned with processes of perception. These processes raise questions about things in the world and the ideas in the mind and the relationships between seer and seen and also about the phenomenology of perception.

DOI: 10.1057/9781137275363

Whilst he is critical of Bergson in many respects, Merleau-Ponty's phenomenology has some features in common with Bergson especially in his later work. This focus upon the interrelationships between mind and body and between perception and what is perceived and subjects and objects is useful in considering the possibilities of thinking about future time which can be expressed as the challenge to 'time's winged chariot', as well as providing a means of understanding how different elements of time fit together. Merleau-Ponty, in his later work (1968) on visibility and invisibility, develops his understanding of the relationship between subject and object in radical ways which depart from his earlier work on the phenomenology of perception (1962). His later work is much closer to Bergson's work on the interrelationship between subject and object and between perception and the inner-world of ideas and the world of objects and materials. Merleau-Ponty describes the crisscrossing of the seer and what is seen and, that is what is visible, between the toucher and the object which is touched thus crossing the boundaries between the senses and between subject and object. In his work, Merleau-Ponty uses the idea of flesh not as something substantive or as self identity, but rather as divergence (écart) or the gap between touching and being touched, that is to say between the sentient and sensible aspects of existence (Merleau-Ponty, 1968) which presents a new way of understanding perception by exploring the relationship between people and things. Flesh in this sense is not based on a privileged, animate category but is the most elementary level of being. Flesh is thus the subject's difference from itself and is reversible because it can fold in on itself with a dual orientation with states which can be re-directed both inward and outward to the world. The to and fro of these processes move the 'real time' of now into the future since through dual orientations both inward and outward applies to the dual orientation between present and future times.

The enfleshed possibilities of future time are concerned with change and continuities in sport in which body practices aim to sustain corporeal powers and achieve endurance. Bodies, which are part of Merleau-Ponty's idea of flesh, age and Marvel was not the first, or the only poet, to note and bodies are subject to the exigencies of time in sport, in the possibilities of extending the barriers and boundaries of generation and impairment, and of infrastructural constraints, including social systems of exclusion such as racialisation, sexualisation and patriarchy.

The Games may be more democratic than many fields of sport but some social divisions and exclusions endure, including those of sex gender.

DOI: 10.1057/9781137275363

Although the 2012 Games were vaunted as the women's games, media times still persist in prioritising men's time. For example, 1 August 2012 was the day of team GB's first gold medal won by Heather Stanning and Helen Glover in the women's pairs. Although there was a mention at the lunchtime news on BBC coverage, by the afternoon when Bradley Wiggins had won gold in the men's individual cycling time trial the two women's achievement was completely forgotten. All BBC news led with Wiggins' gold followed by Michael Jamieson's silver in the men's 200m breaststroke swimming and made no mention at all of the women's pairs.

Men's achievements occupy prime time, for example at the top of the news, and women's are absent. The possibility of men's success always takes precedence even over the actuality of women's. There are the spatial aspects of visibility in media coverage of sport and sporting events but the temporal dimensions are just as important. Men's sporting achievements and the expectation of those possible achievements are prioritised, for example in coming first: the timing of the reporting of the event and the ordering of the communication about the event.

The timescale and the order is what creates and reinstates inequalities. There may be differences in the times achieved by women and by men in comparable competitions but there are also differences in the ordering and prioritising of those events. Differences between women and men in challenging records and achieving the best times remain crucial to the establishment of inequalities: any challenges or subversions are challenged, as in the case of Ye-Shiwen the 16-year-old 200m and 400m individual medley swimmer who swam faster than US men's champion Ryan in the last 50m on 31 July at the 2012 Games and was consequently subjected to accusations of cheating and demands for further drug testing (The Telegraph, 2012).Ye-Shiwen's gold medal swim was described as 'insane' by former record-holder Stephanie Rice who has already denied foul play. 'There is no problem with doping, the Chinese team has a firm policy so there is no problem with that'. Ye won the 200m medley at the world championships last year, but her 400m medley swim which reduced the world record by 7 seconds was surprising for so young an athlete, but Leonard admitted such an improvement was possible at her age (The Telegraph, 2012) .

'Women's Time' is the title of a very influential article by Julia Kristeva in which she situates what she calls the problematic of women like other socio-cultural groups such as young people in time (Kristeva, 1981). She explores the temporal dimensions which, she argues, have been

DOI: 10.1057/9781137275363

under-represented in much of the literature on gender as well as culture more broadly.

> 'Father's time, mother's species,' as Joyce put it; and, indeed, when evoking the name and destiny of women, one thinks more of the space generating and forming the human species than of time, becoming, or history. (Kristeva, 1981: 15)

Kristeva uses the example of European women who are nonetheless related to women across the globe. She uses a particular empirical example but seeks to provide understanding of women's experience by situating women temporally and spatially. This time however is one in which 'the feminist movement both inherits and modifies' (Kristeva, 1981: 15).

Sexual divisions are used to demarcate who can and who cannot participate in competitions, what they are allowed to do and may also shape experience and perception in the relationship between seeing and what is seen which makes experience. Time and the possibilities of futurity are subject to political forces in accounting for which phenomenological accounts lack some of the explanatory force which can locate the operation of power and sources of inequality, which are also subject to and entwined in temporalities. If the future involves some idea of progress, it has to be conceptualised within a theoretical framework which is attentive to inequalities in experience and, most importantly for my purposes in this chapter, to an understanding of how the future might be different and for creative imagining. Sometimes the experience of the present offers some insight into the temporal mobilities and imaginings through which change might be effected and a new future imagined.

Feminist futures

As Elizabeth Grosz argues, there has been only limited work on time which has been undertaken by feminist scholars (Grosz, 2005); although recently there has been research into the connections between temporalities and community, which draws attention to the need to acknowledge contemporary social relations and political issues which have specific temporal dimensions and involve the idea of moving forward and changing in time through local experiments and transforming social relations (Time and Community, 2012). Grosz suggests that Druscilla Cornell's work (Cornell, 1991, 1992) has been important in its contribution to new ways of thinking about time and especially about futures.

DOI: 10.1057/9781137275363

Feminist futures have also connections to the idea of being in the zone, if rarely, using the concept directly. The idea of the zone and the concept of being in the zone do appear in the sports literature especially within the context of social psychology, most notably following from and drawing upon the work of Mihaly Csikszentmihalyi on the focused energies of the experience of flow (Csikszentmihalyi, 1975). The idea applies to experiences of transcendence in many fields at many different levels, like being on a roll, in the groove in jazz or getting your eye in, in cricket or riding the perfect wave in surfing, although the concept was initially developed to explain psychological states rather than transform social relations. One difficulty with a focus on the flows which characterise the internal experience of being in the zone is that it seems to be individual rather than collective, and the political transformative properties of the state remain unharnessed. This is not necessarily the case however. The idea has capacities for moving out of the inner worlds where the state of being in the zone is experienced into the wider social arena where opportunities are created and thus incorporating new ideas about futurity.

Although not directly addressed to the concept of 'being in the zone' in any explicit ways, Cornell's approach does offer insights into how an exploration of timelessness and of transcendence in the present might be linked to theories of temporality and in particular to understanding futurity. What is innovative about Cornell's work, although it is not directly addressed to temporality and certainly, like most feminist philosophical accounts and conceptualisations, does not engage with sport, is her argument supporting the view that any political project which seeks to improve the future must engage with the present and its troubles. Grosz argues that all ethics and politics which are directed at ameliorating the present must be attentive to the future (Grosz, 2005). Cornell's calls into question the more familiar feminist approaches to futurity which have stressed the promise of sci-fi and technoscience in generating utopian futures which, whilst creating interesting and entertaining ideas, are not politically motivated to deal with the actualities of inequalities. Feminist sci-fi has been more involved with a reversal of the present which is then represented as the future. Science fiction may tell us more about the present than the future but its explicit purpose is a future conceptualised in opposition to the present rather than a political project which seeks to envisage and make possible a future based on hope and activity resulting from collective action geared to that end.

DOI: 10.1057/9781137275363

Cornell addresses the issue of becoming, in the Deleuzian sense of movement and of change, whereby the process of becoming is not one of imitation but a removal from the original functions in order to bring about something new (Deleuze and Guattari, 1987). What is significant and productive about the idea of becoming is its mobility, change and flight, which can be marshalled to envision futurity and transformation. Feminists such as Rosi Braidotti (1994, 2002), Clare Colebrook, (2001) and Elizabeth Grosz (2005) have used this idea to propose feminist futures which subvert the patriarchal present to ensure that the future does not resemble the past. Rather than being a reaction to or engagement with the present as science fiction is, Cornell's future is a feminist imaginary domain which is not only imaginary but is virtual, that is new but not yet actualised. (Cornell,1991, 1992). Thus the future is neither preordained nor necessarily constrained. Feminist futures have to be imagined beyond these constraints without falling into the trap of utopianism.

Whilst these approaches draw extensively upon the work of Gilles Deleuze and his collaborator Felix Guattari (Deleuze and Guattari, 1987), they also use more overtly feminist work such as that of Luce Irigaray (Irigaray, 1991,1984). Irigaray's politics of difference has been very influential in challenging the patriarchal logic of Lacanian psychoanalytic thinking, but in the context of rethinking futurity, especially in the empirical context of sport, about which I have to admit none of them has ever written anything, because of her concern with the imaginary and the state of *jouissance* (Irigaray, 1991, 1984). *Jouissance* is state which predates language and Freud's and Lacan's law of the father and the oppressive regime of patriarchy. Irigaray demonstrates the absence of the female pre-oedipal imaginary and of representations of the mother–daughter relationship in Western culture which is represented in Lacanian pyschoanalysis with its denial of female desire and makes the case for a powerful female imaginary which derives from female corporeality. The idea of *jouissance* has found its way into sport as an alternative to the zone since it is used to describe the pleasurable embodied states of either transcendence or harmony which can be experienced through the extremes and routines of physical activity which characterise sport. In the case of the women-rowing pairs, harmony and synchrony were in the sporting activity and *jouissance* in the release of recovery.

The purpose of this consideration of some feminist approaches to futurity, which are not explicitly studies of future time but which nonetheless are primarily engaged with change and the promise of transformation

DOI: 10.1057/9781137275363

in sexual politics, is first to challenge the restrictions of considering the future either as a prediction or as a utopian dream. The idea of becoming embraces mobility and change without presenting a linear trajectory, which is also one of the weaknesses of models of the future as prediction and predictability. Feminist appropriations of becoming also permit change which is not simply set in opposition to the present but could be emergent and creative in genuinely transformative and innovative ways. Second, these approaches offer an alternative to the clock time of measured temporality and offer the possibility of rethinking time not only as social but also as political.

The experience of timelessness can be part of actual enfleshed activity within sport, which has variously been expressed as being in the zone, transcendence or jouissance. Experiencing what is described as timelessness can also be a virtual experience which can be actualised in the present as a route into engaging with the energies and immediacies of now which, in going beyond discourse and beyond the language of description might offer the possibility of future time too. Timelessness in sport generates possibilities for future time and a reconceptualisation of then future through the immediacy of now. *Jouissance*, especially as developed by Irigaray, is a political project as well as a description of a pleasurable state thus facilitating a conceptualisation of futurity which sees time as social and, most importantly for a study of time in sport, as inextricably interwoven with corporeality. Time is enfleshed as well as social.

Future time: transformations and promises

Flesh and the possibilities of transforming bodies in the pursuit of sporting excellence have ever-more purchase in the field of sport and the Olympics, whatever their rhetoric of social inclusion, democratic participation and traces of amateurism, and they demonstrate strongly how future time is construed through the ever faster, stronger body in the discourses of the Olympic Movement as well as in the Olympic Charter (IOC, 2012) and, of course, in the training regimes of athletes and the vast troop of experts who accompany all elite athletes. The Olympic Charter acknowledges some of the social and political possibilities of the games, for example in widening opportunities for participation to those who have been marginalised and discriminated against, but the focus on the future is one which is linear and concerned with prediction and targets;

DOI: 10.1057/9781137275363

a quantifiable rather than a creative imaginary future, which nonetheless is part of what can be so democratising about sport, especially the Olympics.

On 11 August 2012 at 8.30 pm I watch the women's 800m. I am keen to follow the South African 21-year-old Caster Semenya, as I have written about her experiences of gender verification testing (Woodward, 2009, 2012a). In media interviews on the subject I have experienced some of the expressions of prejudice which she has been subjected to. She may have been re-instated since her 11-month suspension following her 2009 World Championships victory but I am aware that her performance has deteriorated since her most unfortunate treatment at the hands of the IAAF, the international regulating body of athletics which conducted the tests, largely on the basis of her outward appearance as a tall, strong woman and the disgruntlement of fellow athletes, which included leaking the results of testing. I want Semenya to win. She starts way back and stays there for the first 400m. It looks as if she is leaving it a bit late to surge past the Russian athlete Mariya Savinova who is well ahead of the field. The race is less exciting than some I have watched but I wonder if I am too conscious of the political issues as I watch to immerse myself in the now of the competition. I am too busy analysing and stepping outside the moment this time. Semenya wins silver with Savinova way ahead to win gold in the fastest women's 800m in 2012, but the commentators say 'Semenya was never in it. Maybe left it too late?' Semenya never had the record, but she was dismissed in the post-race discussion as irrelevant even though she got the silver medal. Her speed has been variable since her disqualification largely for appearing too masculine in 2009. She did qualify with the fastest time in 2012 but is not really seen as a serious contender on this occasion (BBC Semenya, 2012). Maybe the commentators feel uncomfortable. Gender verification testing, even if correctly carried out without leaks to the press and malpractice, is still a troubling approach to classification and the setting of boundaries. Such testing assumes the existence of two distinct sexes as categories of person and brooks no alternatives and no crossing of boundaries (Woodward, 2009, 2012a). It also assumes that with sufficient knowledge and techniques of testing the truth could be found and revealed (Woodward, 2012a). Such tests are based on a linear trajectory, for example, genetic inheritance, birth, whatever life course, notably living as a woman, or maybe passing as a woman and being seen by others as a woman. The IOC even sets time limits for the period after surgical transgender alignment after which a

DOI: 10.1057/9781137275363

person can be deemed to participate in the sport in the gender category to which they have been aligned. The system is based on both historical assumptions and narratives of cheating; ethical concerns, about men trying to pass as women and gain unfair advantage, and historical cases intertwine in what Antonio Gramsci called stratified deposits of common sense (Gramsci, 1971) as well as in the statutes of the governance of sport (IOC, 2012).

Much of the governance of sport is underpinned or possibly haunted by fear of malpractice and cheating. Whilst performance enhancing drugs and the risk of men passing as women and thus gaining unfair advantage have been the main concerns, especially the former with the advent of more sophisticated pharmaceutical interventions, in 2012 some of these issues about what is fair and what is not were raised at the Paralympics.

Paralympic time

It is 2 September and I am watching the 200m, T44 final. Oscar Pistorius the famous Blade Runner is the favourite. He is my favourite too, along with many others who are watching. He is also a big name, an athlete of considerable renown not least because he has participated in the Olympics and Paralympics, in sport for athletes not categorised as having any impairment, even though he has been a double amputee since the age of 11. Watching from a distance as I am, it is not immediately or even at all apparent that these athletes have blades to enable them to run; anyway I forget even though I remember that he is famous and I have the knowledge of his past, having written about him. Pistorius doesn't look as big as some of the other athletes. I wonder briefly how important this is but they are off and Pistorius is ahead at 100m and it looks as if he is going to win. Suddenly, it is sudden because this is a fast race and they are very near the end, the Brazilian Alan Oliveira surges ahead. It's all over and Oliveira has won at 21.45 seconds. Pistorius is second with 21.52 seconds. Pistorius is not taking defeat well. He is complaining and remonstrating about how unfair it is that the Brazilian has longer blades. This time 'incredulity' is applied to injustice and not the athletic achievement for which Pistorius has, quite rightly, been so celebrated. Pistorius says it is 'ridiculous', maybe because he has so surprisingly been dethroned. Pistorius may have failed to qualify in the Olympics heats but only just and he remains the most well-known ambassador and hero of the Paralympic movement.

DOI: 10.1057/9781137275363

Later, I learn he apologises for these outpourings in the heat of the moment. He then went on to win the 400m on 9 September in a blaze of glory. These are the strands of sensation and affect which make up the immediate moment of now in the games, but which in this instance technologies and technical enhancement are visible and addressed in the making of futures. Futures are not only about achievement and faster times, they are also concerned with the ethical problems they pose. The present and how we engage with the present means being attentive to the dilemmas that are thrown up, not least in relation to technological and pharmaceutical enhancement. This is not only a matter of readjusting the categories of disability. Technoscience is advancing, notably in the possibility of merging the prosthesis with the bone to reduce the pain that amputees suffer. The use of prosthetics raises questions about how far sporting futures might go, even to the extent that able-bodied athletes might be able to enhance their performance by modifying their bodies, even amputating limbs to use prostheses to this end. This is not sci-fi, this is a futurity which can be envisioned in the present and beyond the present.

Ethical judgements, technological and scientific advances, economic and social infrastructures are all implicated in the governance and practice of the Olympics, sometimes expressed in the language of social inclusion and progress.

Conclusion

This chapter has suggested some approaches to future time and futurity which do not rely exclusively upon the projection of the present into the future in the form of prediction, although this is an important dimension of future time in sport. By focusing on now and interrogating the sensation of immediacy and immersion in the present, as well as the social context and meaning of temporality, it becomes possible to envisage alternatives which go beyond the present as well as positing alternatives to utopian dreams. The language of sport is embedded in both quantifiable systems of measurement and dreams and aspirations, which are often expressed in hyperbole and traditional excess, but the processes of attempting to capture 'real time' make it possible to deconstruct how 'real time' is made and the promises it offers.

What I have called 'real time' is of course a negotiation between past, present and future and necessarily includes some reflection, drawing on

DOI: 10.1057/9781137275363

past knowledge, as well as projection into the future. Sport offers a route into exploring the future through the present in distinctive ways because of the centrality of enfleshed activity and all the dimensions of future time: the idea of progress and of technological development, prediction and measurement, utopian futures and risk and danger, an assemblage which opens up new possibilities.

There is a dialogue and interrelationship between perception and materiality which is demonstrated by the importance of bodies as explored in relation to flesh and becoming enfleshed. Bodies, their practices and performance are central to the measurement of time. Precision in these processes suggests that it is possible, whereas experience and perception are subject to ambivalence and illusion and challenge some of the claims to certainty of such systems.

The arguments which have been made in this book about the social nature of time and its contextualisation apply to futures too. Because time is social, it is always tied up with social relations and social inequalities. Thus, as Cornell argues, any critique of the present which analyses social relations and puts forward the promise of change and of remedying inequalities must build a conceptualisation of the future into the present, but one which is not constrained by the present. This approach to futurity makes change possible whilst accepting the constraints of time and the inevitability of its passing.

DOI: 10.1057/9781137275363

6
Changing Times and Changing Time

Abstract: *The final chapter uses the example of the games to evaluate the extent to which the regulatory frameworks of sport can create possibilities for change and the promise of transformation as well as reinstating constraints and fixities. This book broadens the scope of analyses of time and temporality by arguing that sport not only reflects existing understanding of temporality but also generates ways of being in time and of conceptualising the processes. In spite of new achievements and records which make sport a constantly changing field there are endurances and entrenched stabilities, but the book has suggested new ways of looking at time and the possibilities of being in time as well as competing against time.*

Woodward, Kath. *Sporting Times*. Basingstoke: Palgrave Macmillan, 2013. DOI: 10.1057/9781137275363.

I started the project of writing this book with the proposition that sport was an under-researched field for the exploration of time and temporality even though time is crucial to sport in so many ways; notably in the setting and keeping of timed records of achievement, especially of moving very fast and also the rehearsing and remembering of sporting stories made through memories of sporting times. I was unsure about what I would find but wanted to use the experiment of writing in 'real time' to try to find out more about time and temporality through the experience of being and writing in 'real time'. There is something about the immediacy and intensity of sport, both of actively participating in its embodied practices and of spectatorship and the attachments of following sport, which lends itself to an exploration of how past, present and future might come together in the moment of being there.

The final chapter of *Sporting Times* considers some of the challenges and opportunities which this short study of time and temporality in sport has presented to enhance the understanding of temporality. The book has covered a number of different dimensions of the social contexts of time and explored a range of ways in which time is relational and cannot be taken for granted, either as fixed or linear. Although sport, especially the Olympics, seems fixated with time especially in the form of records, memories and histories, in this chapter I evaluate the extent to which the regulatory frameworks of sport and its enfleshed practices and virtual properties and presentations can create possibilities and the promise of change as well as reinstating constraints and fixities. In spite of new achievements and records which make sport a constantly changing field there are endurances and entrenched stabilities, some of which present a very unequal playing field, and the book has suggested new ways of looking at time and the possibilities of being in time as well as competing against time. The future is not just a target to be met; the future might hold promises and opportunities too.

Although sport is saturated with records and the measurement of time, whether as specified in the regulations of the duration of a sport or of the speed of completion of an event, of which the fastest must be the 100m. Time and timing are part of the game and the timescales of sport filter into the discourses and common sense of everyday life, saved by the bell or a game of two halves. However, as the book has shown in the context of the Olympics, it is never that simple.

Rules are complex in all sports, not just the Olympics of course. Football may be a game of two halves, specifically a game of 90 minutes

DOI: 10.1057/9781137275363

of two 45 minute halves (FA, 2012) but in football too extra time can be allocated for cup games to resolve a draw. Injury time is added on to compensate for time lost during the game. American football (NFL, 2012) can last much longer. Penalties can go on for a long time if both sides are equal. Although I have demonstrated that in the Olympics the regulatory framework seeks to limit the length of time a game can go on, there are other cases where the length of the game can lend excitement to spectators, if not exhaustion to players. In tennis the final set has to be won by two games and so can go on for a seemingly interminable period of time as it did in June 2010 at Wimbledon in the men's singles between John Isner and Nicholas Mahut, lasting a total of 11 hours and 5 minutes with a fifth set of 8 hours 11 minutes (The Guardian, 2012).

Regulatory frameworks have to negotiate the tension between protecting athletes from exhaustion and possible injury, providing an entertaining display and ensuring both fair play and a competitive edge. Each of these aspects is underpinned by time and timing. Clock time is only one aspect of time in sport, but I have argued that although clock time dominates the moment of participating watching it is not as separate from other aspects of time as might at first appear. Clock time is always enmeshed with other forms of time; not only are the technologies of measuring clock time implicated in social and economic systems but the experience of clock time is also part of the social, cultural and political processes in which progress is made and new systems are developed.

Making sense of sporting time

This exploration of time in sport is instructive about the features of time and about moving towards a better understanding of temporality in a number of ways. The experiment of 'real time' demonstrates some of the ways in which past, present and future which are distinctive characteristics of time interrelate in what remains, of course, the irreversible progress of time as a forward movement. What I have attempted to demonstrate in this book is that this is not a simple trajectory nor one which is unrelated to other forces, including those in play in the wider social and cultural terrain of which sport is a part, as a field in which social relations are made as well as reflected. Given the centrality of bodies, embodiment and flesh in sport, this is a field in which it is productive to explore the relationship between mind and body and between ideas and

DOI: 10.1057/9781137275363

things, the phenomenology of perception and the connections between seer and seen.

I have suggested that a study of time and temporality in sport can offer insights into a wider study of temporality partly because sport offers a field that is distinctive. It is distinctive not only because sport is obsessed with times and timing in the establishment of records and of targets but also because sport and the experience of the Olympics and Paralympics in 2012 in particular show the intensity of how now is translated into the future. Sporting times as illustrated by the Olympics are distinctive because of the mix of emotion and attachment and the immediacy of the moment of 'real' live time, but the games are subject to same infrastructural forces of inequality as other fields of social life and even generate some of their own, especially in relation to sex gender, class, race and disability. In the games, affect binds people together too – affects such as emotion, excitement and engagement as well as aspiration, attachment, hope, and future promise both inside the events and looking forward to the next Olympics and Paralympics in Rio in 2016.

The features of time which emerge from this study are framed by what is the distinctive characteristic of time as an active force. This is the forward march which is irresistible and irreversible, if not as linear and simply defined in terms of progress and change as might have been thought. This is a complex active force, because time is always social and enmeshed with the social forces which are implicated in its forward trajectory.

Human endeavours, especially in sport, seek to harness time through both technologies of measurement and training regimes and other apparatuses to improve performance. Time, however, is not added on to distance or to a separate space; it is not the case that time is one thing and distance another. Time is always part of the mix and underlies other forces, actions and movements. Attempts to intervene in sport, for example, by strategies aimed at improving times are illustrative of another distinguishing feature of time; it does not transform itself. Body practices, record keeping, measurement techniques and myths and legends as well as social and economic infrastructures, all intervene to effect change. Through the regulatory frameworks of sport, such as timed events, records set by best times, precise recording of times, transformation through technological developments, constant search for certainty and boundaries which permeates sport which extend well beyond timed records of sporting achievement to gender verification; categories of sex as well as disability attempt to secure certainties are set by temporal parameters. The social

DOI: 10.1057/9781137275363

and the cultural uses the temporal because time scales and the movement of time is always there.

Time is significant in sport, as this exploration of the 'real time' of the 2012 Olympics shows, because it is the before and after properties of time which make sporting records, practices and stories possible. Stories and legends which make up sporting times are constituted and put into discourse through the passage of time. Writing in 'real time' demonstrates that there is no real-time moment that is outside what has been, for example the collective knowledge which I brought to each event, and the promise of futures which generates the affects of emotion, attachment and excitement and of being part of a wider community which distinguish sport. By focusing on 'real time', some of these components can be disentangled; the connections and the disconnections and disruptions within the assemblage of strands which make each event and the Games themselves, which are always situated within time. This is what makes now so sensational and so spectacular on occasion.

The forward trajectory of time is unfolding and transforming. The Olympics 2012 demonstrated some of the progress that has been made in the history of the modern Games, being marked by the expansion of women's participation and by the enormous success of the Paralympics, within a particular discourse of social inclusion and widening participation. What changes is social and cultural but the very concepts of progress and of transformation are made possible only within the framework of temporality. There are changes and endurances, for example, in the ways in which inequalities and social exclusions play out in sport but the idea of progress can be mis-identified with the forward movement of time. Time passing is inevitable and so are some changes, notably those related to the flesh and to the forces of generation. Bodies' age and are subject to the exigencies of sporting enfleshed practices; technological and medical interventions may have made enormous progress in redressing some of the risks and dangers of sport and in improving longevity but there are barriers that cannot be overcome. However many older athletes there were at the games in 2012, including Hiroshi Hoketsu in the Japanese equestrian dressage team who at age 71 was the oldest participant at the Olympics and Kate Murray of Team GB, aged 64 in the archery at the Paralympics, there are limits to what is or could be possible. It is the social, cultural and political forces which are situated in time and in which time is inextricably tied-up which are what creates change.

DOI: 10.1057/9781137275363

Time is always social because the perception of time is what materialises temporal forces and movements. The rhythms of time and the experience of time passing are phenomenological and embodied and how time is understood affects how changes are forged through political collective action. How change is perceived, for example, as progress or as retrograde is the outcome of the interrelationship between social forces, perception, and systems of temporality. For example, I have argued that women's boxing in 2012 was indicative of progress, but this is strongly disputed and there is a strong case for seeing the participation of women in a sport in which the main purpose is to render one's opponent unconscious and to inflict damage as retrograde. The debate however is located within a context of progress and temporal transformation in the rhythms of time. In order to situate positions in the argument, it has to be related to the passage of time and the ways in which progress and political advances are understood. These issues have emerged from the experience of writing in 'real time' which has reinforced the view that not only is time social, but what is social is also temporal and in sport there has been more attention paid to spatial dimension of its organisation and practices which take time for granted rather than addressing what is distinctive about sporting times.

Change and becoming

One of the features of time in sport which emerged from the experience of being in time and watching the games was the relationship between fixity and finality, and fluidity and mobility in relation to time. On the one hand technologies of measurement of time abound all designed to quantify the passage of time precisely and accurately but this is also a changing field; technologies, practices and even bodies are all changing within time. Watching the games was impossible without participating in the knowledge exchange of records over time and projections into what might be possible. The 100m always attracts debate about how fast it is possible for humans to run (Clouder, 2012). Sport is a fast-moving field, and not just in the context of the making and breaking of timed records. Sports science is a huge growth area (Chang et al., 2001; Page, 2006; Sewell et al., 2012) as is sports psychology. Sports science covers introductory texts, handbooks and guides for athletes and coaches for improved performance. Texts such as Michael Kent's 3rd edition of the

DOI: 10.1057/9781137275363

Dictionary of Sports Science demonstrate the enormous developments in the field to accommodate genetic engineering, biomechanics, fitness testing and training, kinesthetics, sports medicine and nutrition (Kent, 2006).

The approach taken to an investigation of sporting time and times in this book highlights two central and related issues in the exploration of temporality which are also particularly generated by the study of sport. First, by using the trigger of the 2012 Games and an engagement with the present and the process of watching 'live', it becomes possible to see how time is always social and always connected to social, cultural and political forces. These forces can also be identified and in some ways disentangled to suggest, for example, which inequalities are most persistent and how they are temporally located. The researcher cannot stand outside the intensities of watching 'now', and the ways is which time is used and experienced by athletes is always enmeshed in social forces and translated into past and future times, for example through the records to be broken and the possibilities which each new record creates for the future as well, of course as the timing and tactics used in the present, and in this competition.

Second, the experience of time and the relationship between what is seen, in this case between what is seen, and the seer is all part of the process of being in time. Temporality manifests particular relationships, one of which is this relationship between seer and seen, which has specific resonance in the context of my focus on the now of spectatorship. This is a relationship that involves perception and the connections between objects and subjects and between actual and virtual (Grosz, 2005). The framing of 'real time' also invokes sensation which is suggestive of the possibilities of imaginaries that go beyond time in that they are not restricted either to utopian dreams or to predictions of the future based on the present. Sport offers ways of being in time, lost in the moment and the experience of timelessness which creates the possibility of envisioning a future that is not only based on the present but might be envisioned outside it.

'Real time' and the sensation of live sport may be suggestive of the promise of rethinking futurity but sport remains entrenched in the search for certainties. Sporting time and times are both distinctive and representative of the debates and tensions which permeate the study of time and understanding of temporalities, especially in relation to change and the problem of futures in relation to predictions and quantifiable

DOI: 10.1057/9781137275363

outcome. Sports studies and sports science draw upon the explanatory discourses of physical science, for example in developing ways to transform the material accoutrements of sport, including technological and pharmaceutical interventions, medical and psychological sciences as well as equipment and clothing (Kent, 2006). Sporting times demonstrate the tensions between the ways in which the natural and physical sciences (Kent, 2006) conceptualise material change and the psychological, social, cultural or phenomenological which foreground lived experience with different emphases on inner and social worlds, as the source of reliable knowledge rather than material properties. Sport brings together in very particular ways a focus on the promise of enfleshed transformations with the possibilities of psychological commitment which can lead to change and progress.

As Bergson argued, the material focus of the natural and physical sciences and the experiential and cultural dimensions of studies of temporalities which include perception and understanding present crucial elements of the same set of explanations, rather than a tension to be resolved (Bergson, 1896 [1991]) along with matters of agency and structural, material forces (Bergson, (1889 [1969]). Scientific evaluations and understandings of time are themselves social, and the study of time which might be seen as the prerogative of physics is necessarily social and has powerful social affects and underpinnings.

Deleuzian notions of becoming (Deleuze, 1989, Deleuze and Guattari, 1987) have been used in recent years by some feminist writers as was shown in Chapter 5 (Grosz, 2005, Braidotti, 1994, 2002) along with feminist understanding of temporality. 'Together, they provide us with the parameters of the question of time and becoming' (Grosz, 2005: 180). These are not approaches that have been generally, if ever, applied to sport but I have suggested that the notion of becoming does have some purchase, especially to a discussion of futures and the relationship between past, present and future, especially when viewed from the moment of now, which has been my project in this book.

Time, community and belonging

Infrastructures bind people together – nothing binds people together like the Olympics – either in opposition or support, although in 2012 in the 'real time' of the Olympics and Paralympics there was more cohesion

DOI: 10.1057/9781137275363

and support than opposition. London 2012 was marked by the involvement of large numbers of volunteers who assisted visitors and provided a sense of community and belonging in the organisational infrastructures through which attendance at the Games was experienced. Community is often expressed spatially rather than temporally. As Gary Lineker said in the build-up to the Olympics Closing Ceremony, 'some of the stories of 2012 will last with us forever' which was followed by vox pops backed by the song 'Maybe it's because I'm a Londoner', in the context of legacy. The voice-over commentary continued to state that only a Londoner can answer what's great about the Games in an inclusive discourse of belonging that makes community specific to the host city – of 'us Londoners'. The Games are also described as belonging to the host city rather than the host nation, but in exploring community at London 2012 it can be suggested that the event and the manifestations of community and belonging are more about 2012 than about London, or at least as much temporal as spatial (BBC Olympics Closing Ceremony, 2012)

Community is a temporal belonging which the legacy project aims to prolong and the Games may indeed offer strong opportunities for the materialisation of these aims. Community involves routine and iterative activities through which connections are forged (Temporal Belongings, 2012). The engagement with the specificities of the games as well as their intensities and the reach of their emotive affects forge connections and relationships which have the potential to endure. Infrastructures operate temporally; the more forceful the moment of being in time in the Games, the more chance there may be for this moment to extend into the future. 'Real time' is also routine and the Games albeit for a relatively short period of time became everyday, with daily updates, news bulletins which led with medals and outcomes in sporting events with such saturated coverage that routine exchanges even between strangers often involved some reference to a sporting hero. The Games replaced the weather as a topic of conversation.

Regulatory frameworks and bodies include the media which play a huge role in both the commercial sports nexus and in creating and communicating what is social and cultural about sport as well as generating excitement and the psychic investments that go with the whole enterprise of what makes sport matter. The media coverage even manipulates emotion and affect through time as Channel 4's coverage of live events has been used to demonstrate in this book. These strategies are deployed to sustain the psychic investments that are made in the to and fro of

DOI: 10.1057/9781137275363

sensation through spectatorship. Time and emotive affects are entwined in ways that are specific to particular sports too. Time is central to the experience of sensation but is specifically related to embodied sporting practices and varies with different sporting events; cricket has reasonable advertisement breaks at the end of an over or during the tea break, but in the Games viewer interest has to be sustained during the event with the reassurance on your return after the commercial break that the 'live' event picks up where it left off. The promise of the broadcaster is framed by claims to the manipulation and management of time but it is emotion that is manipulated through these techniques. Time and social and cultural forces work together in a sense. Media infrastructures manipulate as well as respond to audiences and in sport are most closely tied into the generation of sensation and emotive affect. It is feelings that enmesh spectators and money that fuels participation.

These processes of generating order and cohesion in the language of community in sport involve regulatory frameworks: notably in the Olympics these were composed of the IOC along with the NOCs and the specific organising committee of the games, which in 2012 were LOCOG as well as the commercial interests, sponsors and agencies of government, all of which are deeply implicated in the assemblage of elements which make up the sporting event. The intersection of power axes in these assemblages includes the categorisation of persons and sporting competitions and body practices which are framed by time which focuses upon futurity and the notion of progress.

Measurement permeates sport, especially the Olympics and Paralympics, with, for example, increasingly complex and precise categories of disability which are carefully classified with the nature of disability along with categories of sport with their letter codes with numerical subcategories such as S for swimming T for track and subdivided by the extent of impairment. The fact that uncertainty in outcomes is intolerable in sport and in the mechanism of classification and measurement, for example, of precise speeds which are integral to its regulatory processes, might suggest that sport is not about fluidity and indeterminacy. These regulatory practices have to be seen in the context of time rather than just being about recording time. The desire for certainty is also part of the human condition and of the underlying irreversibility of time as an active, forward-moving force.

As I have argued clock time is not the only time and 'now' incorporates past and future. Sport offers possibilities and a different kind

DOI: 10.1057/9781137275363

of movement of time which includes the remembering of what is past into the present as well as the promise of transformation which could be future time. Speed, distance, amateur and professional status, and even sex have to be bounded and measured. There is no scope for error and no tolerance of cheating, however prevalent it might be, or of ambiguity, but there are different, temporal versions of fluidity which relate to the future and to change whereby, by seeing time as social, it is possible to imagine the future and changing social relations in the present.

Temporality is social and the experience of time is specific within social contexts, which in no way undermines the irreversibility and forward-moving trajectory of time. The inevitability of the passage of time may have rendered it a taken-for-granted force which is somehow separate from social forces which I have argued it is not. The temporal and the social are inextricably mixed especially in the perception and understanding of temporality and the promise of becoming.

In making a contribution to the understanding of temporality, futurity and the promise of futures have played an important part in the intellectual framing of this book. I am going to end with a beginning which is indicative of some of the ways in which time is understood, if not always directly as social and as an event and an experience which brings together many of the social dimensions of time and temporality with which *Sporting Times* has been concerned. The opening ceremony of the Olympics in 2012 featured a non-linear, idiosyncratic narrative of Britishness which drew strongly on history and accounts of the past; the opening ceremony of the Paralympics created a vision of futurity framed strongly by the forces of science and technology.

Beginnings and endings

On the 29 August 2012 at 8.30 pm BST I watch the Paralympic opening ceremony which is already underpinned by the knowledge that these Paralympics are to be the most ambitious and well-attended yet in the modern games. There have been some tensions prior to the start about the extent to which the Paralympics could replicate the splendour and the spectacular display of the Olympic opening ceremony. I am not disappointed as it starts with the Big Bang and Stephen Hawking. This is the start of time which puts the sporting records of mere mortals into the shade. There is no shade in fact but a lights display to outshine the

DOI: 10.1057/9781137275363

Olympic opening ceremony. This is the past, as far back as we can go at the start of time, but more emphatically this is the future, set within a context of Enlightenment rationalism which challenges the mediocre popular culture of the Olympics closing ceremony party. The future is possible and science provides the promise. It is technology which keeps the physicist Stephen Hawking active in his research in spite of the ravages of motor neurone disease and total paralysis. He speaks with his voice synthesiser which is so familiar in his public performances that it is his voice. There are umbrellas and disabled people under umbrellas. In the background there is popular music with a repetitive beat, which sounds very tedious aurally, but the display is so visually impressive. Able-bodied dancers gyrate about conventionally. This time, the actor Ian McKellan is back as Prospero from the *Tempest* mixing Shakespeare as cultural inheritance and the past into the present with an expression of futurity. A child is suspended in a wheel chair and the journey of exploration begins with an evocation of Newtonian physics represented by an apple, followed by Human Rights and the Higgs Boson in an eclectic mix of social justice and scientific faith and rationality, lit by the startling brightness of scientific hope and truth, which nonetheless transmits a coherent message linking time and the social forces which are implicated in and drive the Olympic and Paralympic movement and demonstrate some of the dynamic processes of the experience of sporting times.

Light is everywhere synthesising science and light with futurity and promise which nonetheless revisits tradition with the royal presence of the Queen and the head of International Paralympics Committee. The Queen's arrival is accompanied by a military march with Union Jack through the stadium resonant of the military links to Stoke Mandeville and soldiers who have been injured in Afghanistan. Past and present come together in this celebration of futurity. The Duchess of Cambridge appears looking serious and in a more grave mood than has been hitherto seen at the Olympics, for example in her enthusiastic support for Andy Murray winning tennis gold.

The National anthem is sung, soberly and quietly with acknowledgement of the impairments of athletes who are participating; some use sign language. There is a conventional parade of athletes very much in the traditional style of the Olympics. The commentary tells us that 'Paralympics are what makes us human – our strengths and our weaknesses'.

This is the beginning not only of the Paralympics but it is an expression of futurity set within a frame of light, the hope of future and belief

DOI: 10.1057/9781137275363

in science. This is the expression of future time in 2012; it is social time and human attempts to engage with time. This feeling of promise was expressed in retrospect at the 10 September street parade celebrations, which brought the greatest show on earth 2012 to a close in a last hurrah for the Games which also drew upon the language of hyperbole, but which nonetheless sought to invoke some of the temporal belongings of a particular moment in time, 'real time' to draw upon the concept of memory projected into the future: 'A moment of visual poetry soon to melt away, unlike the memories that will brighten the bleaker days ahead' (Williams 2012: 3).

The date 10 September 2012 was when London sealed the 2012 memories with 'a victory parade that seals the memories of six weeks of wondrous deeds' (Williams 2012: 1). Futures can be optimistic or pessimistic; they can also be surprising and unpredictable. Sporting times and an immersion in 'real time' offer new ways of envisioning futurity and of exploring how temporality and social forces interrelate with the possibility of looking outside the present into the future. These are the days of miracle and wonder indeed.

DOI: 10.1057/9781137275363

References

B. Adam (2004) *Time* (Cambridge: Polity Press).

B. Adam (1994) 'Perceptions of Time' in T. Ingold (ed.), *Companion Encyclopaedia of Anthropology: Humanity, Culture and Social Life* (London: Routledge).

B. Adam (1995) *Timewatch; The Social Analysis of Time* (Cambridge: Polity Press).

B. Adam (1990) *Time and Social Theory* (Cambridge: Polity Press).

AIBA (2012) http://www.aiba.org/ [last accessed 16 September 2012].

W. Bagehot (1966) *The English Constitution,* 5th edition (London: Fontana).

L. Barnett (1957) *The Universe and Dr Einstein* (New York: William Sloane).

BBC Beijing Closing Ceremony 2008 (2012) http://news.bbc.co.uk/go/em/fr/-/sport1/hi/olympics/7578133.stm [last accessed 3 September 2012].

BBC Fina Swimwear (2012) http://news.bbc.co.uk/sport1/hi/olympic_games/7944084.stm [last accessed].

BBC Great Olympic Moments (2012) http://www.guardian.co.uk/sport/series/50-stunning-olympic-moments [last accessed 15 September 2012].

BBC Olympics Closing Ceremony (2012) http://www.bbc.co.uk/sport/0/olympics/19235488 [last accessed 15 September 2012].

BBC Olympic Stories (2012) http://www.bbc.co.uk/p;rogrammes/p00v9bs4 [last accessed 13 August 2012].

BBC Olympic Stories 100m (2012) http://www.bbc.co.ui/programmes/p00v9bs8 [last accessed 13 August 2012].

DOI: 10.1057/9781137275363

BBC Semenya Silver (2012) http://www.bbc.co.uk/sport/0/olympics/18912902 [last accessed 14 August 2012].
S. de Beauvoir (1987 [1949]) *The Second Sex*, trans. H. Parshley (New York: Vintage Books).
H. Bergson (1977) *The Two Sources of Morality and Religion,* trans. R. Ashley Audra and Cloudesly Brereton (Notre Dame, IN: Notre Dame Press).
H. Bergson (1921) *Mind-Energy*, trans. H. Wildon Carr (London: Macmillan).
H. Bergson (1907 [1983]) *Creative Evolution*, trans. A. Mitchell (Lanham: University Press of America).
H. Bergson (1896 [1991]) *Matter and Memory*, trans. N.M. Paul and W.S. Palmer (New York: Zone Books).
H. Bergson (1889 [1969]) *Time and Free Will*, trans. F.L. Pogson (New York: Harper Collins).
K. Boddy (2008) *Boxing: A Cultural History* (London: Reaktion).
R. Braidotti (2002) *Metamorphoses: Towards a Materialist Theory of Becoming* (Cambridge: Polity Press).
R. Braidotti (1994) *Nomadic Subjects* (London: Routledge).
E. Cashmore (2005) *Making Sense of Sports*, 5th edition (London: Routledge).
S.F. Chang, D. Zhong, and R. Kumar (2001) http://ieeexplore.ieee.org/xpl/login.jsp?tp=&arnumber=990868&url=http%3A%2F%2Fieeexplore.ieee.org%2Fxpls%2Fabs_aa.jsp%3Farnumber%3D990868 [last accessed 20 August 2012].
J. Clouder (2012) http://run-down.com/guests/jc_bhayes.php [last accessed 2 September 2012).
C. Colebrook (2001) *Gilles Deleuze* (London: Routledge).
R.W. Connell (1995) *Masculinities* (Cambridge: Polity)
D. Cornell (1992) *The Philosophy of the Limit* (New York: Routledge).
D. Cornell (1991) *Beyond Accommodation: Ethical Feminism, Deconstruction and the Law* (New York: Routledge).
M. Csikszentmihalyi (1975) *Beyond Boredom and Anxiety: Experiencing Flow in Work and Play* (San Francisco: Jossey Bass).
Dangerous Sports (2012) http://unofficialnetworks.com/2011/05/26/dangerous-sports-world/ [last accessed 15 September 2012].
G. Deleuze (2005) *Francis Bacon*, trans. Daniel W. Smith (London and New York: Continuum).
G. Deleuze (1989) *Cinema 2. The Time-Image*, trans. H. Tomlinson and R. Galeta (Minneapolis: University of Minnesota Press).

DOI: 10.1057/9781137275363

G. Deleuze and F. Guattari (1987) *A Thousand Plateaus: Capitalism and Schizophrenia*, Vol. 2, trans. B. Massumi (Minneapolis: University of Minneapolis Press).
Dreyfus, H.L. (1975) 'Human Temporality' in J.T. Fraser and N. Lawrence (eds.), *The Study of Time II* (New York: Springer Verlag), pp. 150–62.
N. Elias (1992) *Time: An Essay* (Oxford: Blackwell).
FA (2012) http://www.thefa.com/ [last accessed 18 August 2012].
F. Engels ([1845] 2009) *Conditions of the Working Class in England* (London: Penguin).
Forbes (2012) http://www.forbes.com/sites/teresagenaro/2012/06/25/frankel-the-most-valuable-racehorse-ever/ [last accessed 3 September 2012].
M. Foucault (1981) *History of Sexuality, No. 1 An Introduction*, trans. Robert Hurley (Harmondsworth: Penguin).
A. Frank (2011) *About Time* (London: One World Books, Simon Schuster).
Frankel (2012) http://www.lovetheraces.com/features/britain-s-first-100-million-racehorse/ [last accessed 3 September 2012].
R. M. Gale (1978) (ed.) *The Philosophy of Time* (Brighton: Harvester Press).
E. Galeano and M. Fried (1998) *Football, Sun and Shadow: An Emotional History of the World Cup* (London: Fourth Estate).
A. Giddens (1999) *Reith Lecture 2 Risk* (London: BBC).
A. Giddens (1990) *The Consequences of Modernity* (Cambridge: Polity Press).
P. Gilroy (2004) *Postcolonial Melancholia* (New York: Columbia University Press).
R. Giulianotti (1999) *More Than a Game: The Social and Historical Aspects of World Football* (Cambridge: Polity Press).
R.Giulianotti and J. Williams (eds) (1994) *Game Without Frontiers: Football, Modernity, Identity* (Aldershot: Arena).
D. Goldblatt (2007) *The Ball Is Round, A Global History of Football* (Harmondsworth: Penguin).
D. Goldblatt and J. Acton (2012) *How to Watch the Olympics* (London: Profile Books).
C. Gosden (1994) *Social Being and Time* (Oxford: Blackwell).
A. Gramsci (1971) *Selections from the Prison Notebooks* (London: Lawrence and Wishart).
R. Greenslade (2012) http://www.guardian.co.uk/media/greenslade/2012/aug/12-national-newspapers-olympics-2012 [last accessed 16 August 2012].

DOI: 10.1057/9781137275363

E. Grosz (2005) *Time Travels. Feminism Nature Power* (Durham, NC: Duke University Press).

The Guardian http://www.guardian.co.uk/sport/1969/jun/26/wimbledon2003.tennis [last accessed 18 August 2012].

A. Guttmann (2005) *Sport the First Five Millennia* (Amherst and Boston: University of Massachusetts Press).

A. Guttmann (2002) *The Olympics: A History of the Modern Games*, 2nd edition (Champaign, IL: Human Kinetics)

J. Hampton (2008) *The Austerity Olympics, When the Games Came to London* (London: Aurum Press).

J. Hargreaves (1994) *Sporting Females* (London: Routledge).

F. Hegel (1970 [1840]) *The Philosophy of Nature*, trans. M.J. Perry (London: George Allen and Unwin).

Horne, J., Tomlinson, A., Whannel, G. and Woodward, K. (2012) *Understanding Sport* (London: Routledge)

IOC (2012) www.olympic/org/ [last accessed 2 October 2012].

Independent on Sunday (2012) 'Super Saturday', 5 August 2012, p. 1.

L. Irigaray (1991) 'This Sex Which Is Not One' in M. Whitford (ed.), *The Irigaray Reader* (Oxford Basil: Blackwell).

L. Irigaray (1984) *Speculum of the Other Woman*, trans. G. Gill (Ithaca, NY: Cornwell University Press).

S. Kern (1983) *The Culture of Time and Space 1880–1919* (London: Weidenfeld and Nicholson).

M. Kent (2006) *Oxford Dictionary of Sports Science and Medicine* (Oxford: Oxford University Press).

J. Kristeva (1981) 'Women's Time' *Signs*, 7, p. 1.

H.J. Lenskyj (2008) *Olympic Industry Resistance: Challenging Olympic Power and Propaganda* SUNY series on Sport Culture and social Relations (New York: State University of New York Press).

H.J. Lenskyj (2007) *Inside the Olympics: Power Politics and Activism* (New York: State University of New York Press).

H.J. Lenskyj and S. Wagg (2012) *The Palgrave Handbook of Olympic Studies* (Basingstoke: Palgrave Macmillan).

LOCOG (2012) http://www.london2012.com/about-us/the-people-delivering-the-games/locog/ [last accessed 15 September 2012].

London Paralympic 2012 Classification (2012) http://www.london2012.com/paralympics/sports/classification.html [last accessed 15 September 2012].

A. Marvel (1997) *To His Coy Mistress and Other Poems* (London: Dover Publications).
K. Marx (1976 [1867]) *Capital, Volume 1* (Harmondsworth: Penguin).
K. Marx (1973 [1857]) *Grundisse* (Harmondsworth: Penguin).
M.Mauss (1973) 'Techniques of the Body' *Economy and Society*, 2: pp. 70–88.
M.Mauss (1979) 'Body Techniques' trans. B, Brewster *Sociology and Psychology* (London: Routledge and Kegan Paul) pp. 95–123.
G. H. Mead (1932) *The Philosophy of the Present,* A.E. Murphy (ed.), preface by John Dewey (Chicago: Chicago University Press).
M. Merleau-Ponty (1970) *In Praise of Philosophy and Other Essays*, trans. J. Wild, J. Eddie and J. O'Neill (Chicago: North Western University Press).
M. Merleau-Ponty (1968) 'The Intertwining-the Chiasm' in *The Visible and the Invisible*, trans. Alphonso Lingis (Chicago: North Western University).
M. Merleau-Ponty (1964) *Signs*, trans. R.C. McLeary (Chicago: North Western University Press).
M. Merleau-Ponty (1963) *The Primacy of Perception*, trans. James M. Eddie (Evanston: North Western University Press).
M. Merleau-Ponty (1962) *The Phenomenology of Perception*, trans. C. Smith (London: Routledge and Kegan Paul).
K. Mitchell (2009) *Jacob's Beach. The Mob, the Garden and the Golden Age of Boxing* (London: Yellow Jersey).
T. Moi (1999) *What Is a Woman and Other Essays* (Oxford: Oxford University Press).
I. Newton (1995 [1686–7]) *The Principia*, trans. A. Motte (New York: Prometheus Books).
NFL (2012) http://www.nfl.com/ [last accessed 18 August 2012].
J.C. Oates (1987) *On Boxing* (London: Bloomsbury).
J.E.G. Page (2006) *Gold Medal Physics. The Science of Sport* (Baltimore, MD: The John Hopkins University Press).
Paralympics Classification (2012) http://www.london2012.com/mm%5CDocument%5CPublications%5CJoinin%5C01%5C24%5C08%5C10%5CLondon-2012-guide-to-the-paralympic-games.pdf [last accessed 11 September 2012].
A. Powley (2011) *When Boxing Was Boxing. A Nostalgic Look at a Century of Boxing* (Yeovil: Haynes Publishing).

DOI: 10.1057/9781137275363

H. Preuss (2008) *The economics of Staging the Olympics: A Comparison of the Games, 1972–2008* (Cheltenham: Edward Elgar).

M. Proust (1982) *Remembrance of Times Past* (London: Vintage).

P. Ricoeur (2006) *Memory, History, Forgetting,* trans. Kathleen Blamey and David Pellauer (Chicago: University of Chicago Press).

D. Rowe (2008) 'Time and Timelessness in Sport', *Sport in Society: Cultures, Commerce, Media Politics*, 11 (2–3), pp. 146–58.

D. Sewell, P. Watkins and M.Griffin (2012) *Sport and Exercise Science: An Introduction,* 2nd edition (London: Hodder Education).

M. Shallis (1983) *On Time: An Investigation into Scientific Knowledge and Human Experience* (Harmondsworth: Penguin).

R. Sheely and L.E. Bourgeois (1994) *Sports Lab: How Science Has Changed Sports* (New York: Silver Moon Press).

E. Smith (2008) *What Sport Tells Us About Life: Bradman's Revenge, Zidane's Kiss and Other Sporting Lessons* (London: Verso).

K. Stewart (2007) *Ordinary Affects* (Durham, NC: Duke University Press).

J. Sugden and A. Tomlinson (2011) (eds.) *Watching the Olympics: Politics, Power and Representation* (London: Routledge).

Temporal Belongings (2012) http://www.temporalbelongings.org/ [last accessed 15 September 2012].

The Telegraph, Ye Shiwen, Swimming Time (2012) http://www,telegraph.co.uk/sport/olympics/olympicsvideo/9439062/Chinese-swimmer-Ye-Shiwen-IOC-comments-on-doping-test-calls.html [last accessed 14 August 2012].

Time and Community (2012) Temporal Belongings Community 92012 http://www.temporalbelongs.org/hope-and-community-futures.html [last accessed 11 September 2012].

Unequal Time, Olympic Cleaners (2012) http://www.thepetitionsite.com/202/764/899/olympic-cleaners-deserve-better-living-conditions/?zoom=20368722 [last accessed 14 August 2012].

L. Wacquant (2004) *Body and Soul: Notebooks of an Apprentice Boxer* (Oxford: Oxford University Press).

L. Wacquant (1995) 'The Pugilistic Point of View: How Boxers Think About Their Trade', *Theory and Society*, 24 (4) pp. 489–535.

S. Wagg (2004) *British Football and Social Exclusion* (London: Routledge).

M. Weber (1989 [1905]) *The Protestant Ethic and the Spirit of Capitalism*, trans T. Parsons (London: Unwin and Hyman).

DOI: 10.1057/9781137275363

B. Wheaton (2004) *Understanding Lifestyle Sport: Consumption, Identity and Difference* (London: Routledge).

R. Williams (2012) 'The Last Hurrah' *The Guardian,* 11 September 2012, pp. 1–3.

K.Woodward (forthcoming) *Global Boxing* (London: Bloomsbury Academic).

K. Woodward (2012a) *Sex, Power and the Games* (Basingstoke: Palgrave).

K.Woodward (2012b) 'The Culture of Boxing; Sensation and Affect' *Sport in History*, 31 (4), pp. 487–503.

K. Woodward (2011) *The Short Guide to Gender* (Bristol: The Policy Press).

K. Woodward (2009) *Embodied Sporting Practices: Regulating and Regulatory Bodies* (Basingstoke: Palgrave).

K. Woodward (2008) *Hanging Out and Hanging Bout: Insider/Outsider Research in the Field of Boxing Ethnography,* 9 (4), pp. 536–60.

K. Woodward (2006) *Boxing Masculinity and Identity: The 'I' of the Tiger* (London: Routledge).

K. Woodward and S. Woodward (2009) *Why Feminism Matters* (Basingstoke: Palgrave).

W. Wordsworth ([1800] 2006) 'Preface to the Lyrical Ballads' in S. Greenblatt (ed.), *The Norton Anthology of English Literature*, Vol. 2 (New York: W. W. Norton & Company), pp. 141–52.

I.M. Young (2005) *Throwing Like a Girl and Other Essays on Feminist Philosophy and Social Theory* (Oxford: Oxford University Press).

E. Zerubavel (1985) *The Seven Day Cycle: The History and Meaning of the Week* (Chicago: University of Chicago Press).

E. Zerubavel (1981) *Hidden Rhythms: Schedules and Calendars in Social Life* (Chicago: Chicago University Press).

D. Zohar (1983) *Through the Time Barrier* (London: Paladin).

DOI: 10.1057/9781137275363

Index

DOI: 10.1057/9781137275363

DOI: 10.1057/9781137275363

Lightning Source UK Ltd.
Milton Keynes UK
UKOW040605131112

202109UK00002B/3/P

9 781137 275356